THERE'S NOTHING IN THE MIDDLE OF THE ROAD BUT YELLOW STRIPES AND DEAD

ARMADILLOS

THERE'S NOTHING IN THE MIDDLE OF THE ROAD BUT YELLOW STRIPES AND DEAD ARMADILLOS

JIM HIGHTOWER

HarperPerennial
A Division of HarperCollins*Publishers*

A hardcover edition of this book was published in 1997 by HarperCollins Publishers.

HarperCollins books may be purchased for educational, business, or sales promotional use. For information please write: Special Markets Department, HarperCollins Publishers, Inc., 10 East 53rd Street, New York, NY 10022.

First HarperPerennial edition published 1998.

Designed by Nancy Singer

This book is manufactured in the U.S. on totally chlorine-free paper made by Lyons Falls Pulp & Paper, and printed with soy-based ink.

The Library of Congress has catalogued the hardcover edition as follows:

Hightower, Jim, 1943–
There's nothing in the middle of the road but yellow stripes and dead armadillos / by Jim Hightower. — 1st ed.
p. cm.
Includes index.
ISBN 0–06–018766–2
1. United States—Politics and government—1993– —Humor. 2. United States—Social conditions—1980– —Humor. 3. American wit and humor. I. Title.
E885.H56 1997
973.929—dc21 97-23211

ISBN 0-06-092949-9 (pbk.)

02 ❖/RRD 10

To
William Fletcher Hightower

Speak the truth, but ride a fast horse.

Old Cowboy Saying

CONTENTS

ACKNOWLEDGMENTS

Public acknowledgments can be dicey and even dangerous, since you almost always leave someone out, thereby leaving them PO'd. At a big political function I emceed in Dallas, I tried to acknowledge everyone in the room who was a candidate for some office or another, from U.S. Senate to county hide inspector. Unfortunately, I left out a guy who was running for the Texas legislature. Not being too tightly wrapped, he already believed there was an intergalactic conspiracy afoot to deny him his rightful place in this august body, and my faux pas convinced him that I was the diabolical mastermind behind the conspiracy. He hounded me for weeks and never forgave me. Worse, he got elected.

So let me say here that I thank *everyone* for this book.

Of course, no one writes a book. *One* does get to put one's name on it, but hundreds of friends, coworkers, acquaintances, writers, librarians, relatives, English teachers, activists, and others have their prints all over the writing. I'll not attempt to list you, but thank you.

There is one special one though: Thank you, thank you, thank you, thank you, thank you, thank you . . . and many, many more thank-yous to **Susan DeMarco.**

A big thanks, too, to my other coworkers at Saddle-Burr Productions, especially to Betsy Moon and Cheri Nightingale, as well as to my agent George Greenfield and to my editor and Big Apple guide Adrian Zackheim, along with his associates Framji Minwalla and Timothy Farrell—all folks who were key to making the book possible.

Jim Hightower
The Chat & Chew
Austin, Texas

INTRODUCTION

We are writing the constitution of a single global economy.
—Renato Ruggiero, 1996

We? Who is we? A new constitution—that's big, right? Will we get to vote? Will there be yard signs? Who the hell is this guy, and is he serious?

Serious as a snake bite. Ruggiero, an Italian trade bureaucrat, was hand-picked in 1995 to head an obscure and downright secretive international body called the World Trade Organization, based in Geneva. Technically, the WTO is the creation of governments, including ours, but in fact it is the baby of the world's corporate and financial giants—such names as GM, Chase Manhattan, Nestlé, Caterpillar, Credit Suisse, ADM, Kodak, Nippon, Goldman Sachs, GE, Boeing, Unilever, Exxon, Monsanto, British Petroleum, and, well, the full club of powerhouse corporations that now call themselves "transnational."

For the past few years, members of the club have been diligently separating themselves from allegiance to any national flag, polity, or people, purposefully erecting a legal fortress around themselves, their hoards of money, and their financial ambitions. This legal construct is the "constitution" Ruggiero referred to. It is not merely protection that they seek from their fortress, but a secure place from which to reign, for they are quietly setting themselves up as a rogue "government"—one that is unelected, unaccountable, for-profit, supranational, and sovereign. The building blocks of their "constitutional" fortress are the WTO, GATT, IMF, APEC, OPIC, NAFTA, MFN, CBI, AFTA, MAI, and other jumbles of letters that form the esoterica of global trade policy.

If Paul Revere were to make his midnight ride in 1998, it is not an invasion by Redcoats he would warn us about, but the astonishing assault these global corporate powers are making on our liberties, economic fortunes, way of life, and sovereignty. I'm a Texan without even a pony, so I can't gallop into the night like Revere, but I have written this book as my own modern-day warning, and as a rallying cry to fight back.

What the British monarchy of old could not accomplish by force—the subjugation of the American people—today's new oligarchy of corporations is achieving by stealth, without even a protest from our political leaders. To the contrary, both Bill Clinton's Democrats and Newt Gingrich's Republicans are blithely opening the doors for the transnationals, shouting: "Give me liberty . . . or give me a PAC contribution!"

As I write this, another national election is under way for the control of Congress, yet there is a dumbfounding silence from the whole system about the unbridled power of top corporate and financial executives, who—among their many other excesses—are almost jubilantly knocking down the wages and middle-class aspirations of America's workaday majority. They are abandoning families and whole communities here as they flee to exploit the impoverished in Third World nations. Instead of discussing these real, practical, and immediate class issues—as ordinary folks do every day around their kitchen tables, at their cafes, and in their bars of choice—we get a political season of silliness: Clinton is darting here and there around the globe desperately trying to look like a president; Newt's do-nothing Congress is either in recess or deliberating such weighty matters as renaming Washington's airport for Ronald Reagan; and the media have little time for any politics not tied directly to Clinton's willy.

Far from serious political or journalistic questioning (much less criticism) of their frontal assault on the poor and the middle class, the corporate powers are being widely exalted, hailed as crusaders who are bringing a God-ordained, market-based order to the world's people. They are marching onward under the golden banner of "globalization," a nowhere land in which honey is promised to flow to all if only the owners and managers of transnational corporations are freed from any governmental ordinance anywhere—national, state, or local—that they believe constitutes an infringement on their Holy Right to do business as they see fit. Their self-serving concept of globalization gives their own narrow interest (nothing more noble, by the way, than hauling off the highest possible profit they can grab) supremacy over all competing interests around the world, including those of labor, human rights, environment, religion, democracy, community, and nation.

Such supremacy, for example, can outlaw even a simple effort by your city government to buy made-in-the-U.S.A. products, or to refuse to buy any products made with sweatshop labor, or to choose to buy from hometown suppliers. Imperious trade tribunals already have been established with international authority to strike down these acts of local sovereignty as an infringement on the global trade rights of corporations. Too wacky to be believed, you say? Sony, Toyota, Mitsubishi, and other transnationals are presently pursuing just such an action through the World Trade Organization against the state of Massachusetts, which passed a law in 1997 saying state agencies there will not buy products from companies that do business with the brutally repressive dictatorial gangsters now ruling Burma. The corporations say uh-uh—their global trade rights supersede the legislative autonomy of the Massachusetts people. The case is pending before a WTO tribunal of faceless officials in Geneva. Their proceedings are not open to the public or the press; a state has no standing under WTO rules, so Massachusetts cannot present its own case or even be present; and there is no appeal of the tribunal's ruling.

Having the WTO is better than a 900 number for the chieftains of transnationals, who can dial up 900–SMOOCH-ME and have their every corporate fantasy come true. Chiquita banana, for example, owned by fat-cat political contributor Carl Lindner of Cincinnati, is using the WTO against European nations that now choose to buy their bananas from their former Caribbean colonies. He says they must change their laws and buy bananas from his operation in Central America, even though such a shift could gut the economies of St. Lucia and other small island-nations dependent on European sales—and even though it would likely shift many desperate islanders into the cocaine and heroin trade. Talk about bananas! Oil companies, too, have already been to the WTO and succeeded in overturning a portion of our Clean Air Act, because its antipollution standards were ruled too strict and therefore a barrier to the dirtier oil products the companies ship to the United States from Venezuela.

The rationale for this absolute and absolutely absurd ascendancy of corporate interest is that ultimately their pursuit of private gain will produce untold global good by softening the hearts of dictators, making capitalists of child laborers, generating a worldwide consumer society based on exports, and giving all of us pie in the sky when we die.

I was born at night, but it wasn't *last night*; how about you? This is nothing but Ronald Reagan's "trickle-down" nonsense writ large—much larger than he and his laugh-a-minute "Laffer curve" lunatics ever dared attempt. It is the codification, too, of George Bush's gibberish about a "New World Order," which he was too inept to implement. But then along came Bill Clinton, a sheep in wolf's clothing, a "New" Democrat who is unabashedly obsequious to gray-templed corporate executives. He led the Democratic Party—once the proud political home of working folks—into an unholy alliance with Republicans and Wall Street lobbyists, and the next thing you know we were being hit by NAFTA, GATT, WTO, AFTA, CBI, MAI, and other initials that, if you scramble them, spell "gotcha" in all the Romance languages. They sell these deals by using the high-minded rhetoric of Free Trade! Jobs! Exports! Growth! Prosperity! and other come-ons that twinkle and wink at us like stars in the dark sky, appearing as though we could almost reach out and grasp them. But their high-mindedness is a deceit and the benefits chimerical. Every one of these trade agreements and structures rips more economic and political power out of our hands and conveys it on a silver platter to the elite club of international speculators, powerhouse bankers, corporate *jefes*, and other hustlers—men with smiling lips and squinting hearts.

Amarillo Slim, the legendary poker player, once offered this useful tip: "Look around the table. If you don't see a sucker, get up, because you're the sucker."

Look around America's table. There sit the Democratic and Republican parties side by side, not a dime's worth of difference between them, waving us voters to the table once again for congressional and presidential elections, while their corporate contributors keep shuffling the deck and grinning slyly at us—can you win anything in their political game? Maybe you are among the 80 percent of Americans who don't like the economic hand you've been dealt, whose incomes are flat or even falling. Hey pal, say the pundits and politicos fronting for the powers-that-be (whose stack of chips is so high you can't see their faces), our economy is humming, so there must be something wrong with you. You've still got a chance, though, they tell this workaday majority—a "new millennium" is coming, so ante up, try to cross that "bridge to the twenty-first century" and become a "winner." And look, there's Bill Clinton himself, presiding over the whole casino with his arms spread wide and his face beaming, proclaiming over and over, "These are good times for America."

How surreal. In Clinton's "good times" incantation one hears the time-warped echo of Herbert Hoover, who tried back in 1928 to sucker the people with his insistence that Republican prosperity had "placed the whole nation in the silk stocking class." The problem with such gloating presidential pronouncements is that they are patently goofy, since most folks simply are not experiencing what the Big Man is claiming for them. People naturally ask, for example: If we really were in the silk stocking class, wouldn't we be wearing silk stockings?

Likewise today, when the Tub Thumpers of the Economic Orthodoxy (including not only the Democratic President, but also the GOP's daffy duo of Newt & Trent, as well as the entire coifed chorus of network news sparklies) go on and on about the "boom" in America, most folks ask: A boom for whom? Their situation is similar to that of West Texas dryland farmers watching a line of thunderclouds form on the far horizon and praying for even a small rain shower on their crops. They see bolts of lightning in the distance (Flash: Michael Eisner's 1997 pay as CEO of Disney Inc. was $575 million. Flash: Wall Street brokers are so flush with bonus money this year, they are ordering $10,000 bottles of wine with lunch. Flash: Bill Gates built a $40 million new house containing slightly more square footage than Rhode Island). Then they feel the rumble of the thunder rolling across their land (Boom: Corporate profits up! Boom: Dow Jones Average up! Boom: Worker productivity up!). It's a spectacular show, but as West Texas farmers have learned the hard way, thunder ain't rain.

Thunderous boosterism about good times might comfort the moneyed elite, but it does not fool the folks, certainly not the eight out of ten of our fellow citizens living in the dryland economy. In the essays that follow, I write from the viewpoint of these people, the everyday working folks I grew up among in Texas, the people I campaigned with during my "politician period" and championed while in state office, the people I now encounter through my political travels around the country and on my national call-in radio show, *Hightower's Chat & Chew*.

My experiences out here—outside the power centers of money, media, and politics—tell me that our nation is in a very dangerous period, rapidly approaching a crisis. Dangerous because the people of the *United* States are becoming fundamentally *dis*united—not because of abortion or any of the other widely reported disputes over social issues, but because of raw economics. The disunity is widening as the superrich, the merely rich, and the affluent (roughly the wealthiest 20 percent of American families, but especially those in the very top tier of privilege) aggressively and openly separate themselves from the economic fortunes and goodwill of the many.

Much more than economic stratification and social hierarchy is in play here, however. Those at the top are also abandoning America's basic principle of economic fairness, tearing asunder the powerful, unifying belief within our diverse population that *we're all in this together*. Let me repeat: In this time of highly advertised prosperity, eight out of ten Americans are falling behind economically. Nor is there any prospect of their position improving. Indeed, there is not even official acknowledgment, much less a policy discussion, of their situation, for these same people also are being shut out politically by the two major parties, which have put a platinum price tag on being a player with them. When the political ante is a thousand dollars, the 80 percent majority has no place at the table.

Here's the joker in the deck for those in charge: Where is this majority to go, and what are they to do? Is it imagined that so many will sit still for such treatment very long? Bear in mind that, for the most part, we're not talking about poor people. Our leaders are well practiced at holding down the poorest one-fifth of our population, labeling them "welfare cases," riffraff or worse, and hiring ever more police with ever more police power to contain them in their own neighborhoods or in prisons. But we're talking now about *four-fifths* of the American people, the true middle class as well as the poor—the factory workers, back-office clerks, teachers, carpenters, small farmers, nurses, plumbers, janitors, shopkeepers, data processors, and all the others who sweat, nurture, plow, invent, instruct, build, arrange, and otherwise make our country work, sustaining it from the ground up. These are folks who earn maybe $20,000-$50,000 a year, folks with expectations, aspirations, and attitudes. They know about the productivity gains U.S. companies are enjoying in the nineties, the unprecedented new wealth that has been created, the corporate profits and stock values that have soared—because they helped produce all this. Yet practically all the economic benefits generated by the many have been forklifted to the top, leaving them stewing in the realization that their families' middle-class opportunities are being stiffed.

In the following pages, I am rude enough to call this what it is: class war. A devastating class war is raging in America's countryside. The corporate media don't cover it, and you can bet that candidates of both major parties will studiously avoid any mention of it in this fall's congressional campaigns, because this is not a class confrontation in which the many are menacing the wealthy few—a turn of events that would definitely make the nightly news, complete with fulminating politicians calling forth police to beat back the aggressors. Rather, today's class assault has been initiated from the top—a ruthless war of attrition in which the wealthy few, from their bastions of privilege on Wall Street and in Washington, are lobbing bomb after bomb into the lives and communities of the many.

"We are too quiet," I heard John Lewis say in his quiet but forceful voice to a crowd of attentive people standing around him. "We need to make some noise." Lewis is the Atlanta congressman who was side by side with Martin Luther King Jr. in the bloody-ugly confrontations of the civil rights movement in the sixties, but on this day in Columbus, Georgia, February 1998, he was urging on a downtrodden group in another struggle, not against the brutishness of Bull Connor racism, but against the brutishness of global corporatism.

Lewis is part of a dissident bunch of Democrats who are making periodic forays into the countryside this year to meet informally—personally—with everyday Americans to learn about the local realities of global theories. Representative David Bonior of Michigan heads the group, which was organized by the Ralph Nader–backed Citizen Trade Campaign, and I was invited to come along for the ride and broadcast my radio show from the road, which I joyfully did. These are hardly your typical, lobbyist-financed congressional outings—we wend our way from urban to suburban areas, from small towns to farms, traveling aboard a groaning old bus, chowing down on cold sandwiches and potato salad for our lunch and supper, and staying overnight in the spare bedrooms of local supporters. Especially impressive is that these loquacious lawmakers do precious little speechifying and a whole lot of listening and note taking on our voyages. The people do not hesitate to give them an earful.

Outside the gates of a modern but now-closed plant on the suburban edge of Atlanta, Anna Harris talked with the lawmakers about how she and a thousand people like her had been kicked off the edge of the global economy by Lucent Technologies, a $26-billion-a-year giant that makes telephones and other products. For twenty-five years she worked for Lucent, gradually working her way into America's middle class, earning $15.59 an hour, or roughly $31,000 a year. She and her coworkers were highly skilled, efficient, cooperative, and loyal workers. Quality. Just like the telephones they made.

Still, Lucent's executives constantly messed with them, saying Mexico beckoned. Ms. Harris and the others were told in the early nineties to take a pay cut . . . or else. They did: "I went back to $13 an hour," she told the lawmakers. "I'm a single parent. I took a cut in pay to keep my job." That reduction sliced $5,000 a year from her middle-class life.

Even this giveback did not appease the wanderlust of Lucent, however. Shortly after Clinton and Congress rammed NAFTA into law in 1993, the corporate honchos hitched up the wagons in Atlanta and hauled Anna Harris's job to Reynosa, Mexico. In this poverty-stricken border town, only a stone's throw across the river from Texas, the company pays its Mexican hires a subpoverty wage of $1 an hour. They get no benefits, unless you count the daily ration of one breakfast taco supervisors hand out to each worker. Lucent uses their labor; then, thanks to NAFTA, the company merrily trucks the Mexican-made phones right across the bridge into the U.S.A., delivering them to a store near you without paying any tariff or honoring any quota.

Now in her early fifties, Ms. Harris finally landed another job months after being abruptly abandoned by the globally wayward Lucent. She got one of those "14 million new jobs" that a crowing Bill

Clinton tells us his economic policies have created. Hers is at a Target store, working for $7.50 an hour, though it's only part-time, so she has fallen plumb out of the middle class back into poverty. Ironically, in her job at Target, she sells the telephones she once made. When Ohio Congresswoman Marcy Kaptur asked her if the price is any lower on those phones now that Lucent pays $1 an hour to Mexican workers rather than the $13 she earned, Anna Harris's eyes turned steel cold and she said: "There's no difference in price. They're selling them for $80 to $90."

Welcome to the New World Order.

—Jim Hightower, May 1998

1

CORPORATEWORLD!

THEY GET THE GOLDMINE, WE GET THE SHAFT

- ★ LOGOS GALORE
- ★ FORE!
- ★ NPM MEMO
- ★ GETTING A LEG UP ON CORPORATIONS
- ★ VERNON JORDAN'S DREAM
- ★ HIGH-TECH EXCESS
- ★ HAPPINESS

LOGOS GALORE

They tell me advice books sell, so here goes:

- Don't ever buy a pit bull from a one-armed man.
- Never sign *nothin'* by neon.
- Always drink upstream from the herd.

Oh, and one more: Never, ever believe the "conventional wisdom," which is to wisdom what "near beer" is to beer. Only not as close.

This is especially true when it comes to our interlocked political and economic systems, in which the game is about power, and conventional wisdom is a trick play designed to keep you in your place. Today's powers-that-be, for example, loudly broadcast that there is no such thing as "class war" in our society, that the major media outlets deliver the "news" to us, that environmentalism is an extraneous concern fomented by liberal "elites," and that we Americans are overwhelmingly a middle-of-the-road, "conservative" people.

These purveyors of conventional wisdom are putting out more baloney than Oscar Mayer, hoping to keep America's political debate from focusing on an insidious new "ism" that has crept into our lives: corporatism. Few politicians, pundits, economists, or other officially sanctioned mouthpieces for what passes as public debate in our country want to touch the topic, but—as ordinary folks have learned from daily encounters—the corporation has gotten way too big for its britches, intruding into every aspect of our lives and altering by private fiat how we live.

Less than a decade ago, for example, your medical needs from birth to death rested in the hands of a doctor, whom you chose. Quicker than a hog eats supper, though, America's health-care system—including your personal doc—has been swallowed damn-near whole by a handful of national corporate mutants called HMOs, most of which are tentacles of Prudential, Travelers, and other insurance giants.

When did we vote on this? Did I miss the national referendum in which we decided that remote corporate executives with an army of bean counters would displace my hand-picked doctor, and would decide which (if any) hospital I can enter, how long I can stay, what specialists I can consult, and what (if anything) these medical professionals are allowed to tell me about my own medical needs? I know Congress did not authorize this fundamental shift to health maintenance organizations (a phrase, by the way, that sounds as warm and welcoming as a lube and body shop). To the contrary, Congress in 1994 rightly trashed the Clinton health-care reform legislation on the grounds that it would do the exact thing being done to us now: limit the choice of doctors and put the bean counters in charge. Only back then we were warned by the infamous "Harry and Louise" television spots that it would be *government* bean counters managing our health.

What irony. For years the very companies that financed the "Harry and Louise" ads have flapped their arms wildly to scare us about that old bugaboo, "socialized" medicine, but while we were looking over there, they blindsided us with something even harsher: corporatized medicine, a brave new world in which the Hippocratic Oath has been displaced by the bottom-line ethos of HMO profiteers like Richard Scott. A mergers and acquisitions lawyer who never cared for a patient in his life, Scott headed the $20-billion-a-year Columbia/HCA corporation until July 1997. A far-flung HMO, Columbia/HCA demanded that its local hospital executives return a 20 percent annual profit to headquarters, or else. How did they meet Scott's demand? By cutting back on services, on employees, and ultimately on us patients.

One place Scott did not cut back, however, was on his own paycheck. In 1995, he took a 43 percent salary hike, which meant he drew a million bucks from the till. A month.

He is far less generous when assessing his industry's responsibility to meet the needs of America's sick, lame, and infirm: "Do we have an obligation to provide health care for everybody?" this captain of corporatized medicine recently asked rhetorically. "Where do we draw the line? Is any fast-food restaurant obligated to feed everyone who shows up?"

While it is true that the corporation has long been an economic powerhouse, since the 1970s it has metamorphosed into something different and disquieting. Just as HMOs have seized control of our health-care system and unilaterally redefined its ethical underpinnings, so have corporations become the governing force in our society, reshaping American life to fit nothing more enlightened or enriching than the short-term profit agendas of these privileged economic entities. Everything from our amusements to our government, from the food supply to the money supply, from language to public education, from the public dialogue to popular culture—all corporatized.

In the summer of 1976, while I was visiting with my folks back in Denison, Texas, my daddy and I were passing a slow Saturday afternoon together, chatting some, catching bits of the baseball game on TV. He was cranked back in his La-Z-Boy and we were sipping a couple of cool ones when an ad for a national chain of fast-food fish houses suddenly blared at us from the television, boasting that "Our fish doesn't taste fishy!" Daddy blinked a couple of times, turned to me, and said, "I like my fish to taste fishy," realizing as he said it that corporations are not beyond tampering even with life's small pleasures.

The true symbol of today's America is no longer Old Glory, but the Corporate Logo. No space, no matter how public the edifice or how sacrosanct, is free from the threat of having "Mountain Dew," "Banc One," or "Nike" plastered on it. Not even space itself is off-limits—one visionary business enterprise is already working on launching a low-trajectory satellite with what amounts to an extraterrestrial billboard attached. It will be programmed to beam various product logos back to earth from the night sky. No matter where you live, from Boston to Bora Bora, you can gaze into the vast darkness, as humans have for thousands of years, and absorb the natural wonder of the moon, the Milky Way, and, yes, an orbiting ad for Mylanta. Lovely.

If the sky can be corporatized, why not our children's schools? Kiddos are a huge market—*elementary* school kids alone spend $15 billion a year from their own allowances, and their parents fork out another $120 billion buying all kinds of stuff for them, from snacks and sneakers to such essentials as hair mousse for first-graders and chocolate-scented perfume by Givenchy. All this before the teen years, when hormonally powered consumerism kicks in.

If you think only television hawks these products to our little nippers, you have not been to a schoolhouse recently. Ads abound, as Jack in the Box, Nike, Coca-Cola, and other corporations rush to take advantage of public-funding cutbacks by offering to help make up a school district's shortfall. All they ask in exchange is a little space on school buses, in the hallways, in the library or gym, on book covers, on the basketball scoreboard, on the school building itself, and anywhere else students might see their ads and begin learning brand loyalty. What is being sold here, of course, is not space, but access to something many parents feel a school board has no right to sell: access to children.

More recently, entire schools have been branded, as various makers of soft drinks, sneakers, computers and other wares bid against their competitors to become the "official product" of Bogus T. Malarkey Memorial Junior High or some such in various school districts across the land. This corporate intrusion caused quite a national flapdoodle in March 1998, when the muckety-mucks of Coca-Cola came to Greenbriar High in Evans, Georgia, for "Coke Day." As part of this product promotion, the cola's marketing executives were invited to address economics students, chemistry students were assigned to analyze the sugar content of Coke, and a Coca-Cola cake recipe was used in the home ec class. To top it all off, Greenbriar students were to line up in formation to spell out "Coke," so a promotional photograph could be taken.

Kids do the darnedest things, though, and senior Mike Cameron had the temerity and sense of humor (as well as the old-fashioned, 100 percent American rebellious spirit) to show up for the photo op wearing a Pepsi shirt. Pandemonium! This is disruptive, wailed school officials, this is insulting to Coke, this is ruining the picture, this is . . . well, it's hurting our chance to get cola cash! For exercising his First Amendment rights and spoiling the corporate day, Mike got suspended from school and sent home. Cooler heads later prevailed, though, and Mike's suspension was dropped from his record, but not before students learned that the corporation is a powerhouse in today's schoolhouse, and it's dangerous to dis the Coke Man.

In case kids miss the corporate handwriting on the walls, advertisers go straight to the classroom through the mushrooming practice of providing textbooks, films, videos, computer software, and other "instructional" materials that also (get ready to be surprised!)

promote the company's products or image. Writer David Shenk has reported in *Harper's Magazine* that one of the most blatant of these corporate come-ons is the ersatz lesson plan and demonstration kit on the "scientific method," distributed free to 12,000 schools across the country by Campbell Soup Company.

Ostensibly designed to introduce youngsters to the ways of science, the kit invites students to conduct an exciting experiment called the "Slotted Spoon Test." What does it test? Whether Prego spaghetti sauce is indeed thicker than Ragú. The lesson materials from Campbell, which just happens to manufacture Prego, put this question to the class: "Have you seen the Prego television commercials that use slotted spoons to compare the thickness of two sauces? It uses steps very similar to the ones you'll use in the experiment." How sciency! Campbell's instructional kit includes a glossy how-to poster, a slotted spoon, and a coupon good for a big jar of Prego. Presumably the class is on its own to get the Ragú, but really there is no need, since the budding scientists are instructed that they have not done the test correctly unless Prego wins the thickness test.

Campbell's soup is hardly alone in using a lesson plan as a Trojan horse to get inside the classroom and inside the heads of children. Hunt-Wesson provides a "Kernels of Knowledge" history package that lists renowned scientists "who made a difference"—Louis Pasteur, George Washington Carver, and . . . Orville Redenbacher? Orville's popcorn is, of course, a product of Hunt-Wesson. Exxon is also in high schools with a video teaching that the *Exxon Valdez* oil spill was not really harmful; chocolate candy companies have a board game in fourth, fifth, and sixth grades teaching that Thomas Jefferson himself had spoken about "the superiority of chocolate for both health and nourishment"; Revlon sponsors a lesson in self-esteem, asking students to imagine "good hair days" and "bad hair days"; education about skin comes to kids courtesy of Clearasil; there is a math "curriculum" that teaches children to count by using Domino's pepperoni pieces; and some classrooms employ a special software that teaches reading by getting the kiddos to recognize the logos of Kmart, McDonald's, Hi-C, and Cap'n Crunch.

These are among some 350 corporations that are disguising their product pitches as "lessons" that reach millions of students. This is a terrific gimmick for advertisers because, as one packager of these in-class promotions tells its corporate clients, "Coming from school, all these materials carry an extra measure of credibility that gives your message added weight." Fine for them, but for the larger society, this commercialization of educational materials insidiously perverts the mission of our public schools from one of teaching citizenship to one of teaching consumerism. Citizenship promotes an activist public; consumerism promotes a passive one—and there you have the deeper corporate interest in dominating and redefining the public education system.

Far from fighting this shift, the country's bought-and-sold political leaders are embracing it, even calling for corporate chieftains to step in and set national standards for our public schools. It received scant media attention, but in March 1996 a National Education Summit was held for the nation's governors and top corporate executives at the IBM conference center in Palisades, New York. President Clinton showed up at this exclusive tête-à-tête to do a bit of corporate cheerleading: "I believe this meeting will prove historic," he asserted, and well it might if the President of the United States so meekly surrenders what should be a public decision to these narrow special interests. Clinton noted that "somebody" must devise new standards for the nation's schools and hailed IBM, AT&T, Boeing, Kodak, Procter & Gamble, and others in attendance for stepping up to do the job—never mind that the corporate agenda for education is one-dimensional, falling far short of what a national agenda should be. Stamping out a uniform line of good workers is hardly the same as preparing young people to become good citizens, which requires, among other things, nurturing a healthy dose of skepticism toward corporate uniformity.

What can you expect though from politicians who are wholly corporatized themselves, dependent on receiving the financial approval of these same Fortune 500 rulers and totally accepting of the conceit that all things public are improved by putting a corporate mark on them. Hell, why not just privatize the government itself?

- Turn the Lincoln Memorial over to Lincoln-Ford so the place can start turning a buck.
- The State of the Union address would seem a natural advertising opportunity for the "Big Whopper."

- Let Barnes & Noble turn the Library of Congress into one of its superstores, putting a Starbucks latte express in the grand foyer leading to the reading room.
- Must the Supreme Court justices wear those awful black robes? Surely Ralph Lauren's Polo clothiers could design something more nineties.
- The Capitol building itself begs refurbishment as a Disneyland operation (LobbyWorld!) with Pluto, Goofy, and the rest of the gang entertaining us alongside Newt and his scamps.

On April 1, 1996, Taco Bell tested these turbulent waters by purchasing full-page ads in the *New York Times* and elsewhere announcing: "Taco Bell Buys the Liberty Bell." Proclaiming that this was a move to help with the national debt, the ad said that the precious American artifact would "now be called the 'Taco Liberty Bell,'" adding that "While some may find this controversial, we hope our move will prompt other corporations to take similar action to do their part to reduce the country's debt."

It turned out to be just an April Fools' joke intended to create a nationwide media buzz around the corporate name and to launch a new $200-million ad campaign for the "world's largest Mexican fast food restaurant." But, in the company's own words, "thousands of calls pummeled the Taco Bell switchboard," and talk-radio callers went ballistic. Interesting. Not so long ago the idea of a corporation buying a piece of America's history and spirit would have been too preposterous for a stunt like this to work, but today . . .

Indeed, the National Park Service itself is pushing legislation to create a corporate sponsorship program that would authorize ten or so "cultural resource partners" to advertise their association with our national treasures. In exchange for donating around $10 million each to the NPS, Black & Decker chainsaws could be the sponsor of Sequoia National Forest, McDonald's could run a TV ad in which Utah's Natural Bridges National Monument metamorphoses into Golden Arches, and Maytag washing machines could buy the right to use the image of the Old Faithful geyser. Too wacky to be seriously considered? Mattel Toys already is discussing the idea of buying the rights to market a special Ranger Barbie—a perky Barbie doll all dressed up in a Park Service uniform. Wonder if it will be made in the U.S.A.? No Barbies are.

Both the Clinton Administration and the Republican leadership of Congress are enthusiastic about the potential for corporatizing these public resources, not only because it fits with the privatization mantra chanted by both parties, but also because it makes them appear to be doing something about our deteriorating network of 374 national parks. But raising a hundred million dollars from corporate sponsors is like trying to fight off a tornado with an umbrella—the backlog of unfunded repairs and other needs within existing parks now surpasses $8 billion. What is needed is not a tacky tag sale of America's natural wonders, but some guts in Washington to provide the public funding needed to preserve these public treasures for future generations, just as generations before us did their part.

Which brings me to the ultimate logo opportunity for corporations: politicians themselves. After all, corporations own them, why not brand them? Have you ever seen a NASCAR race, where the cars, the drivers, the racetrack, the pit crews, and every other inch of space is plastered with ads? A single driver's jumpsuit will have more than a dozen logos on it, from Gatorade to Clabber Girl baking powder, Band-Aid to Lovable brand brassieres, AC sparkplugs to Zippo Lighters, Mr. Goodwrench to Goodyear—and the driver's car is a phantasmagoria of more decals.

I say we should apply this human billboarding to office seekers. In the interest of full disclosure, wouldn't it have been helpful in the last presidential election to see Bob Dole sporting an R. J. Reynolds crest on his blazer, a purple-and-orange FedEx hankie in his breast pocket, the Archer-Daniels-Midland monogram on his shirt cuffs, a Gallo wine belt buckle, a Citibank gimme cap, the Chiquita banana seal on his shirt collar, and a club tie dotted with the logos of his other major sponsors? Bill Clinton, too, could have been corporately accessorized by Goldman Sachs, Time Warner, Tyson chicken, AT&T, Philip Morris, Disney, the Lippo Group, and other brands that favored him. There also should be a rule in Congress that any member speaking for a bill, either in committee or on the floor, has to wear a windbreaker with the name of the sponsoring company stitched across the back and the company logo displayed prominently, if not proudly, on the front.

FORE!

Richard Lederer is the former schoolteacher turned author who has published compilations of off-the-wall answers students give to exam questions. Lederer swears he does not make these up, including this response to a world history quiz: "Gutenberg invented the Bible, Sir Walter Raleigh invented cigarettes, and Sir Francis Drake circumcised the world with a 100-foot clipper."

Ouch.

The corporation has become the Sir Francis Drake of the world of sports, circumscribing this sphere of our lives with its particular set of values and changing the way and the reason the game is played. Indeed, nowhere has the corporate impression on our culture (and on our children) become more visible than in organized sports.

OK, the corporatization of athletics is not a life-and-death issue, but in a society that measures a city's big-league status by its major-league franchises, and that has more citizens watching the Super Bowl than it does voting in the presidential election, the sports scene is not an inconsequential indicator of what is happening in the larger culture.

Check out golf (a game described by Mark Twain as "a good walk spoiled"). The professional golf tour has always struck me as the ultimate expression of the corporate presence in sports, not only because the tournaments themselves are unabashedly organized to hype a company ("AT&T Pebble Beach Pro-Am," "Buick Invitational," "Honda Classic," "MasterCard Colonial," "Lincoln-Mercury Kapalua International," "Shell Houston Open," and "Anderson Consulting World Championship of Golf," to name just a few), and not only because golfers are walking sandwich boards (with corporate logos adorning their shirts, shoes, hats, wristbands, knickers, golf bags, club covers, golf balls, caddies, and every other salable space), but especially because the PGA's ranking of top players is determined each year not by the scores they make, but by how much money they make—Wall Street's true measure of one's worthiness.

If you harbor any doubt that the corporate identity has thoroughly consumed the wide, wide world of sports, just give a listen to

Shaquille O'Neal, the basketball star and primo ad pitchman. In 1996, after signing a $121 million contract with the L.A. Lakers, he was besieged by the media asking about the preposterousness of that sum. Shaq retorted: "I'm tired of hearing about money, money, money, money, money. I just want to play the game, drink Pepsi and wear Reebok."

It is encouraging to note that this corporate takeover is not without its fumbles and stumbles. The total branding of the '96 Olympics in Atlanta, for example, was so garish that there was a nearly universal upchucking by world athletes, spectators, and media alike—not exactly the image Atlanta hoped to project. Still, corporate sponsors were oblivious, relentlessly pushing their presence at every opportunity, including in the so-called news coverage by NBC. In one of the network's syrupy "Olympic Moments" that featured precious-to-the-max profiles of assorted Olympians, Russian weight-lifter Andrei Chermerkin was used to portray the worldwide triumph of corporate ideology over communism. This "moment" showed Chermerkin in the McDonald's franchise in Moscow, cheerfully chomping into a McBurger and presumably delighted that "civilization" had at last entered even into the capital of the old Red Empire. "*Uzhasno!*" he exults. Only after the Olympics were over and the piece had been seen by millions worldwide was it learned by network and McDonald's officials that *Uzhasno* means: "This is terrible!"

Despite such setbacks, the corporate beat goes on—and is growing louder and louder, drowning out the games themselves. Companies that used to advertise their wares on broadcasts of sporting events now are becoming the event. Disney Inc., for example, not only advertises on sports shows, not only sponsors the events themselves ("Disney World of Golf Classic," among others), not only is a broadcaster of sporting events (it owns both ABC and ESPN), and not only has its own sport palaces (it has built the thirty-two-acre, indoor-outdoor Disney World Sports Complex that will host various teams, including spring training for the Atlanta Braves), but it also owns Anaheim's Mighty Ducks of the National Hockey League, owns a fourth of the California Angels baseball team (and will eventually own 100 percent, once present owner Gene Autry takes his final ride into the sunset), and reportedly is maneuvering to buy the Los Angeles Clippers pro basketball team. This would leave Disney only a football team shy of achieving the first Ownership Grand Slam in sports . . . and that could be coming soon.

A group of pro-football executives have had discussions with various corporations and CBS Television about creating a new football league that would have twelve teams—each one representing not a city, but a corporation. Among those named as possible owners are Anheuser-Busch, PepsiCo, and Federal Express. I think this is a terrific idea if for no other reason than it presents the opportunity for some absolutely great team names! Instead of "Bears," "Panthers," and other ferocious animals, why not some more-descriptive monikers that truly reflect the fearsome power of the team owners: the Exxon Oil Spillers, for example, is a natural; the FedEx Unionbusters rings true; the McDonald's Minimum Wagers says it all; millions would turn out to boo the Prudential Policy Cancelers; maybe the GM Job Punters could play in Flint, Michigan; the Monsanto Cancer Causers would strike terror coast to coast; the Lockheed-Martin Cost Overrunners is a winner; and what could be more fitting than the Archer-Daniels-Midland Price Fixers?

THE OLD COLLEGE TRY

Besides hangovers and black-eyed peas for good luck, New Year's Day is a time for the festive pageantry of the college football bowls! Actually, the day has been extended to a couple of weeks, and the pageantry of place—the Sugar, the Citrus, the Cotton, the Gator, the Alamo, the Orange, the Fiesta, the Aloha, the Sun—has been usurped by the celebration of the commercial: the Nokia Sugar, the CompUSA Citrus, the Southwestern Bell Cotton, the Toyota Gator, the Builders Square Alamo, the FedEx Orange, the Tostitos Fiesta, the Jeep Aloha, and the Norwest Bank Sun—and in all cases the company gets top billing. Indeed, some have abandoned any pretense of place or tradition, going with straight corporate identity—the CarQuest Bowl and the Outback Steakhouse Bowl, for example (why no logical and memorable alignments, I wonder, like the Campbell Soup Bowl or the Ti-D-Bol Bowl?).

Bowl games are the least of it, however, as colleges put their entire athletic programs, their facilities, their names, and who-knows-what-next up for sale. Ohio State now plays basketball in the Value City arena. Not that the retail chain built the place, but it simply made enough of a donation to buy the name of the arena.

But the superstar buyers of college athletics are the shoe hawkers. Reebok has UCLA, University of Texas, and University of Wisconsin, for example; Converse has its list of client schools, too, as do Adidas and, of course, Nike—the Gold Medal "Champeen of da Woild" in logo proliferation and corporatization of sports. It is next to impossible to turn on a televised sporting event without being assaulted by Nike's checkmark "Swoosh" logo, which is emblazoned on players, coaches, scoreboards, and every other surface in camera's view. It is so ubiquitous, such a glaring symbol of the regimenting and overbearing rule of money over sport that I see it as less a "Swoosh" than a "Swooshtika."

Nike buys so many college programs—Colorado, Florida State, Illinois, Miami, Michigan, North Carolina, Penn State, and Southern Cal, to name a prominent few—that it has a "manager of NCAA relations," headquartered in Kansas City, just a short ride down the street from the offices of this official ruling body of college athletics. Nike recently even tried to pay the NCAA to let it create (and profit from) a national college football championship playoff series. Putting a brand name on the national championship, however, was too much even for these money-hungry overseers of collegiate sports to swallow . . . yet. Another day, another game.

A short decade ago it would have been unthinkable for reputable administrators of public institutions of higher learning to slap the logo of a notorious sweatshop profiteer like Nike on their "student athletes" and their university, but today it is commonplace. "Just do it," as the Nike ad instructs.

In 1995 Mark Asher reported in the *Washington Post* that for about $1.3 million a year in cash, products, and services, Nike owns the rights to the athletic programs of the University of Michigan. As part of the deal, the company provides the shoes, uniforms, wristbands, gloves, caps, jackets, and other apparel for the players and coaches of all twenty-three U of M teams, as well as uniforms for the cheerleaders and shoes for the band.

What does Nike buy for its money and goods?

1. Its "Swooshtika" appears on all the above, giving the company instant brand-name identity with one of the largest, most successful, and most often televised college sports programs.
2. A commitment that Michigan's coaches and staff will wear Nike gear and will advise all the school's athletes "of the advantages" of Nike products.
3. Promotional appearances by the coaches.
4. Annual participation by the men's basketball team in a Nike-sponsored tournament.
5. Coaches' participation in testing of Nike products.
6. Nike signs on the marquees of both the basketball and hockey arenas.
7. Full-page color ads in the school's game programs.
8. Nike logo on the university's sports publications and videos.
9. Free retail space in the stadiums and arenas on game days.
10. Right to design a new logo for the school and to use it on sweatshirts, caps, and other products for sale.
11. Right to be the exclusive national marketer of these University of Michigan products.

In essence, this public university has hired out as a billboard for the company, corporatizing its name and reputation.

In 1996, Reebok tackled academic freedom itself at the University of Wisconsin. Going for the gold, university officials signed away the U of W name and logo to Reebok in exchange for shoes for its athletes, despite some campus opposition to dealing with a company known for paying slave wages to Third World workers. As in all Faustian deals, the devil was in the details, including one small-print Satan that prohibited campus "disparagement" of the company. The contract declared that the "university will not issue any official statement that disparages Reebok," and the "university will promptly take all reasonable steps necessary to address any remark by any university employee, including a coach, that disparages Reebok."

"Whoa there," screamed faculty and students alike, "flag on the play, intentional foul, push 'em back, push 'em back, waaaay back!" Under pressure from those who do not want to put academic freedom and free speech up for sale to any old shoe company with a duffel bag full of bucks, the Wisconsin officials had to back down and renegotiate, striking the offending language from the contract.

The danger of corporate intrusion is rarely as glaring as in the Reebok case, posing instead a more subtle but equally threatening shift in values from public purpose to private profit. If the bottom line is your only line, then Nike is your model:

- This is a company whose name is used as a verb in Vietnam, where very young women and girls are employed at sub-poverty wages to make shoes the company then peddles here for $175 a pair; to keep the women and girls in line, supervisors find it felicitous to strike them occasionally with one of the company's trademark shoes, thus coining the new Vietnamese term "to Nike" someone, giving a whole new meaning to that "Swoosh" logo.
- This is a company whose honcho, Phil Knight, concedes that he rooted for Brazil to defeat the U.S. team in last year's World Soccer Cup game because—and I quote—"It was a Nike team. America was Adidas."
- This is a company that, right after the '96 Olympic Games, bared its true sporting spirit by airing an ad that proclaimed: "You don't win silver, you lose gold," an avaricious attitude that would make Ivan Boesky, Michael Milkin, Leona Helmsley, and Donnie Trump wet their pants with envy, just as it disgusts such silver medal winners as U.S. swimmer Amy White, who said of the ad: "I am insulted every time I see it. They are slapping every athlete and every country that doesn't win gold in the face. That's not the spirit of the Olympics."
- This is a company that entered a fake team of cross-country skiing competitors in the '98 Winter Olympics in Japan for the sole purpose of getting corporate publicity. Nike recruited two runners from Kenya (where cross-country skiing is unknown), shipped them to Finland for several weeks, and paid a reported $200,000 to bivouac and coach them in the sport's basics. The company then had the two entered as "Kenya's team" in cross-country skiing competition in Nagano. Kenyan Phillip Boit was pushed by Nike into the

grueling ten-kilometer race. He was not competitive, of course, but Nike didn't give a damn about such Olympic ideals as individual competitiveness; it was after a marketing score. It got what it wanted when Boit finally crossed the finish line, dead last. The Norwegian winner of the race happened to be there, too, and he reached out in an international gesture of sportsmanship and shook Boit's hand. What a PR windfall for Nike! This touching "Olympic moment" was broadcast to a worldwide television audience, which saw a tired but smiling Kenyan completing the course, the very personification of the "I can" ads that Nike was running nonstop during the Games. Not by accident, Boit was swathed in corporate "Swooshtikas," just in case the audience missed the crude advertising connection. As sports columnist Bob Woznowski so vehemently put it: "These [two Nike entrants] are not athletes clearing hurdles to reach their Olympic dream. These are marketing pawns financed by well-heeled publicity seekers." Nike itself retorted with less than Olympian idealism: "People forget, we are a business, and part of our objective as a business is to get attention."

THE ART OF THE DEAL

In high school I lettered in a couple of sports, but I never could cut it in basketball, perhaps because I was short, slow, and clueless. I did make the team as a second-stringer back in the ninth grade, although nearly all my court time was restricted to practice, giving the A-teamers someone to run around.

It was in a practice session late in the season that I finally realized I did not have a future in this game. I was dribbling the ball down court to the right when I made a pretty good basketball move, whirling around to my left to drive daredevilishly for the basket . . . only to crash head-on with our team's star, Harold Ballou. His forehead caught me right in the mouth, and we both sprawled on the floor. He was fine, only a little abrasion, but his thick skull had snapped off my front tooth and turned my mouth into a bloody, mushy mess. Yet all the trainers, coaches, and other players rushed to Harold, encircling him, patting his hand, and cooing solicitously at him, leaving me on the side by myself, spitting blood and dazedly groping around on all fours for my lost tooth, which it turns out I had swallowed.

All these years later I find an allegory for our times in this junior-high collision: Harold-the-Star represents the swaggering owners and other corporate powers that control today's sports racket; the solicitous coaches and trainers are our state and local politicians; and that dazed guy bleeding all over the court is you, the taxpayer.

That the sports world is being wholly corporatized is only the second-most underreported story in sports; the *most* underreported one is that you and I are subsidizing the corporatization. While every sport literally is awash in statistics, with our daily newspapers devoting at least a full page of diminutive, six-point type to chronicling up-to-the-minute scores, standings, and averages (rivaling, aptly enough, the amount of type devoted to daily stock market stats)—the one sports statistic that is never reported is the running total of taxpayer subsidies flowing to those elites sitting in the sky boxes. As Casey Stengel often declared, "You could look it up."

Those corporate-sponsored golf tournaments and bowl games, those Nike "Swooshtikas" festooned across every stadium, those donations that put a company's name on a college gym, those companies that bought a piece of the Olympics in order to be designated the "Official Reconstituted Meatlike Product" or whatever of the show in Atlanta, and all the other corporate imprints are being written off by those corporations as promotional, advertising, and "charitable" expenditures—corporate tax loopholes that let them quietly slip their multibillion-dollar sponsorship tab to us.

Then there is the old-fashioned, direct government handout, and no one has their hands out more than the big shots who own the franchises, mostly demanding that cities, counties, and states pony up from the public till to build them shiny new stadiums and arenas with enough chrome, upholstery, and design gewgaws on them to make a '57 pink Cadillac look plainer than an old brown Plymouth.

Every sport is playing this game. Among the gabillionaires presently receiving or seeking public alms to build luxury houses for their private enterprises are Ross Perot Jr. (Dallas Mavericks basketball), Microsoft's Paul Allen—the eighth richest man on Planet Earth (Seattle Mariners baseball), media baron Tom Hicks (Dallas Stars hockey), and Eddie DeBartolo Jr. (San Francisco 49ers football).

In the September 6, 1996, issue of *USA Today*, Erik Brady and Debbie Howlett wrote an excellent six-page special on the "Stadium Binge," which opened with this on-target characterization: "Remember the rich kid on the block who always threatened to take his ball and go home if he didn't get his way? He's been replaced by the rich owner who vows to take his ball club and *leave* home if he doesn't get a new place to play." To keep the team in town, taxpayers are asked to build a facility to the specifications of the team owner, the owner leases said facility for a pittance (or for nothing), the owner gets to keep all the revenue on everything from ticket sales to the beer concession, and the owner can walk away from the facility whenever the mood strikes or a better deal comes along. There even are deals that promise to tax local citizens and give the money to the owner if the team's revenues fall beneath a certain level. Shouldn't these guys have to wear ski masks when they walk into city hall?

Still, politicians are eager to appease big business, and they fear being blamed for losing their city's "major-league status," so they almost always go along with the blackmail. Invariably, these politicians, the chamber of commerce, and the local media dutifully hail the *jobs!* created by subsidizing the team owner, but nearly every independent economist agrees that this is hogwash, that taxpayers always lose, pointing out that most of the payroll hailed by officials goes to players who do not live there in the off-season, and to fat-cat executives and relatives hired by the owner, with local working stiffs lucky to get a few part-time, seasonal, minimum-wage jobs.

Despite this, the boom is on! North Dakota Senator Byron Dorgan, no fan of this taxpayer thievery, notes that "stadiums are the only healthy public housing program we have left in this country." Brady and Howlett found that at least forty-five professional sports facilities were being built around the country at a cost of more than $9 billion, with four out of five of those dollars coming not from the owners who milk the profits from these cash cows, but from you and me—fans and nonfans alike. Worse, the cost of attending a game in these Taj Mahals has increased to the point that the families of ordinary taxpayers have been shut out.

Owners don't regret shutting out us "bleacher bums," however, because they find upscale customers more desirable for their publicly financed facilities. They are eager to bulldoze their existing sports houses, even if these are wonderful places to watch games, and beloved sites that help define a city's character, because of one thing: sky boxes—or, as the corporate marketers prefer to call them, "luxury suites." Read the stories about each of the forty-five facilities to be completed by 2000. The common complaint from each owner was not that the existing field or floor where the players perform was inadequate, or that there were too few fans at games, but that there were either not enough or no luxury suites for preferred clients. "So build me a new place laden with sky boxes," barks the owner to the mayor, "or I'll find some politicians in another town who know how to deal."

The depths of this absurdity have yet to be plumbed, but surely Dade County, Florida, approached the bottom. Only ten years ago, the county completed a pink, art deco-ish, $53-million hall to showcase both the Miami Heat NBA franchise and the city's NHL franchise, the Panthers. But Wayne Huizenga, the baron of Blockbuster Video who owns the Heat, and Mickey Arison, the cruise industry magnate who owns the Panthers, grew restive: *"Not enough luxury boxes!"*

"Yes sir," replied the local establishment, and both pouting owners are now getting new, taxpayer-funded arenas—a $212-million hockey hall for Arison and a $210-million basketball chateau for Huizenga, each with plenty of princely suites they can lease at upward of $100,000 a pop. Meanwhile, the perfectly fine Miami Arena has gone from pink sports palace to white elephant in less than a decade, and local taxpayers are left holding the bag for $38 million in bonds still owed on the abandoned arena.

We lose more than tax dollars in these rip-offs. A seismic cultural shift also has taken place, akin to the proliferation of gated-and-guarded living enclaves for the privileged: sky boxes represent the economic segregation of the "stands," one of the few remaining social settings where Americans mingle democratically, where the banker might sit next to the beautician and the cheers of the CEO mix with those of the laborer. Sky boxes, however, now lift the few above the public, carving out exclusive, glassed-in spaces of conspicuous elitism right in the middle of the populist rabble.

The Picasso of these high-flying stadium deals is the artful dodger himself, Art Modell, presently the proprietor of the Baltimore Ravens football team . . . by way of Cleveland.

This guy wins the "Dirty Play of the Decade Award," hands down. For years Modell was the lucky owner of one of the most successful NFL franchises in history, the Cleveland Browns. In this shot-and-beer, working-class city hard up against Lake Erie, the fans were not merely loyal to the Browns, but true fanatics, with the diehards in the end-zone bleachers so raucous and rowdy that the space was proudly known as the "Dawg Pound"—football at its primal best, with every home game a citywide civic event.

The city turned out an average of 70,000 ticket buyers a game and produced the highest TV ratings of any NFL team. Such fan loyalty made Modell, a former daytime television producer, wealthy beyond his wildest dreams. By 1995 he was the majority stockholder in a football franchise valued at more than $160 million. Like all team owners, though, Modell claimed to be just a civic-minded businessman barely keeping afloat, poormouthing that it *cost* him money to bring football to Cleveland's burghers. *Financial World* magazine, however, pegged the Modell team's annual profit at $8 million, not counting whatever side deals, fat salaries, and executive perks he might have taken for himself.

Not enough, apparently. With a wandering eye, Art slyly detected a chance to make a bigger killing in Baltimore, which had lost its own team to another corporate pirate in 1984 and had been begging the NFL hierarchy ever since to get a replacement. Art smelled sky boxes. But his dream of reaping riches in Baltimore seemed dashed in the early nineties, when the NFL announced it was going to expand by two teams, and the Maryland city was the leading contender for one of the slots. According to *New York Times* sportswriter Thomas George, at a closed meeting of NFL owners to consider each city's bid, one owner vociferously and successfully opposed an expansion team going to Baltimore, passionately insisting that the NFL had no business expanding there and that pro football could not make it in the city, even "banging his fist on the table for emphasis." That owner was Art Modell.

Two years later he hijacked his own team to Baltimore. In 1995, while Cleveland fans continued faithfully to pack the Dawg Pound for every Sunday's game, Art was secretly slipping in and out of Maryland to cut a deal that would rip the Browns out of the hands of Cleveland fans. This guy has all the characteristics of a dog, except loyalty. He finally sealed the deal under the cover of darkness on October 27, 1995, meeting with Maryland Governor Parris Glendening aboard a private jet on the tarmac of Baltimore's airport.

And what a deal it apparently was. Despite public money being spent, details of these payoffs are rarely made public, but various sources have reported that Modell is getting a $200-million new stadium (with 108 sky boxes!—plus 7,500 premium-priced "club seats") built for him by the state and leased to him *rent-free;* he collects the revenues from all concession sales and parking; he gets a $15-million training complex; the state has renovated an interim stadium to his specifications while he awaits completion of his new digs; he received up to $75 million for moving expenses; he gets a $50 million cash payment just for doing the deal; and taxpayers will guarantee ten years of sellouts for the new stadium.

How to pay for this giveaway? Governor Glendening and the legislature took money out of state lottery revenues, imposed a new tax, and tapped other state funds. "Wait a minute," protested a few legislators, who thought maybe the school kids of the state might have a more pressing need for these public moneys than a runaway multimillionaire from Cleveland. But the avarice of Art rose to the challenge: "I feel sorry for their position," he said of the rebellious legislators. "I feel for the schools," he said, just warming up, "but look at the positive effect of pro football on a community, the emotional investment of people at large. You can't equate that with fixing up the schools."

THE BOY SCOUT

Debonair dance man Fred Astaire once observed: "The hardest job kids face today is learning good manners without seeing any." In today's corporate climate, good manners are the least of it. Kids see basic values like honesty, loyalty, fair dealing, respect, trustworthiness, responsibility, and team play (each a virtue ostensibly taught by sports) routinely stiff-armed by the likes of Philip Knight and Art Modell. By their ethical standards, the bottom line is the goal line, and one does whatever it takes to run up the score.

This is a lesson seeping deeper into our culture than we might wish. This year's top seller of Boy Scout popcorn, for example, is a thirteen-year-old in Georgia who devised a popcorn-selling strategy much slicker than the time-honored tradition of going door-to-door or setting up a table at the shopping center. Instead, according to a piece in the *Atlanta Journal*, this son of a corporate tax accountant said to himself: "Wait a minute, this is Boy Scouts—there's got to be a tax deduction."

With his father's help, the lad put together a flier and a sales rap that he pitched to companies, asking each of them to buy a hundred dollars' worth of the packaged popcorn to distribute *as gifts*. Clever. If the companies bought the treats to eat, they would get only a $68 tax write-off for each $100 package, subtracting the actual value of the popcorn from the charitable donation. But by doling out the packages to employees and customers as corporate gifts, the popcorn becomes 100 percent deductible.

Bottom Line: As a result of this tax twist, the boy sold $12,000 worth of the product—25 percent more than the previous year's top sales boy. "Basically," this Boy Scout told the newspaper, "the government pays for it. And businesses just loved it."

GO PACKERS!

There is another way. The Green Bay way.

The real champs of the 1997 Super Bowl–winning Green Bay Packers are not Brett Favre, Desmond Howard, and Reggie White—terrific as they are on the field—but the team's owners.

The Packers are totally unique in the NFL. Instead of being owned by Greedheads like Art Modell, this team is owned by "Cheeseheads"—the *people* of Green Bay themselves. With yellow foam "cheese" wedges atop their heads and faces painted with the team's yellow-and-green colors, the diehard Packer fans might appear to outsiders as a bunch of North Country yahoos, but any given one of them could be a Packer proprietor. Some 1,900 of the locals, including truckers, barkeeps, merchants, and bus drivers, own a piece of the Pack, organized back in 1923 as a community-owned, nonprofit company. The stockholders draw no profit, and the locally elected board of directors that operates the team is unpaid, but all concerned draw great pleasure from knowing that the Packers are *theirs*.

What a difference ownership makes. Not a dime needs to be spent to hype up fan support, since the team literally belongs to them. The town of 96,000 built Lambeau Stadium, owns it, operates it, and fills each of the 60,790 seats in it for every home game—forty straight years of sellouts, whether the team is winning or not, and the season ticket waiting list has 20,000 names on it.

Get this: No ticket costs more than $28, no parking space is more than $7, there is free parking within four blocks of the stadium, and on those blustery, bitterly cold, snowy game days, homeowners let fans park in their garages.

The Packers, which began as the team of the Acme Packing Company, is a total community enterprise. Charities run the stadium's concessions, bringing in volunteers who cook the bratwurst and pour the beer, earning some $400,000 a year for their groups. Off-duty police provide stadium security, and are paid overtime by the team. If an overnight storm dumps snow into the seats and aisles, locals simply show up at Lambeau Stadium early with their own snow shovels and other tools and—without being asked, much less paid—clear the snow before the crowds arrive.

Green Bay fans and citizens never have to worry that some pirate of an owner is going to hijack the Pack and haul their team to Los Angeles or any other big-city market, because Green Bay *is* the team. It stands as a shining model of how fans in other cities could get control of their teams and stop corporate rip-offs.

This is why the other owners in the National Football League have teamed up to pass a rule specifically banning any future NFL team from being community-owned.

NPM MEMO

On November 19, 1996, the following internal memorandum was posted on the Internet, apparently by unnamed disgruntled employees of NorthPole Multinational Inc.

WE DELIVER

MEMORANDUM
To: All Employees
From: VP/Personnel
Re: Downsizing

The recent announcement that Donner and Blitzen have elected to take the early Reindeer Retirement Package has triggered a good deal of concern about other restructuring decisions here at NorthPole Multinational.

The reindeer downsizing was made possible through the acquisition of a late-model Japanese sled for the CEO's annual trip. Improved productivity from Dasher and Dancer, who summered at Harvard Business School, is anticipated and should take up the slack with no discernible loss of service. Reduction in reindeer will also lessen airborne environmental emissions for which NPM has been cited and received unfavorable press.

I am pleased to inform you that Rudolph's role will not be disturbed. Tradition still counts for something at NorthPole Multinational Inc. Management denies, in the strongest possible language, the earlier leak that Rudolph's nose got that way not from the cold but from substance abuse. Calling Rudolph "a lush who never did pull his share of the load" was an unfortunate comment made by one of Santa's helpers and taken out of context at a time of year when he was under stress.

As a further restructuring, today's global challenges require NorthPole Multinational to look for better, more competitive steps. Effective immediately, the following economy measures are to take place in the "Twelve Days of Christmas" subsidiary:

The *Partridge* will be retained, but the *Pear Tree* never turned out to be the cash crop forecasted, so we must cut our losses. It will be replaced by a plastic hanging plant, providing considerable savings in maintenance.

The *Two Turtle Doves* represent a redundancy that is simply not cost effective. In addition, their romance during working hours could not be condoned. The positions are therefore eliminated.

The *Three French Hens* will remain intact through next season, although a cheaper, Mexican alternative is being investigated.

The *Four Calling Birds* were replaced by an automated voice mail system, with a call-waiting option. An analysis is under way by Internal Security to determine who the birds have been calling, how often, and how long they talked.

The *Five Golden Rings* have been put on hold by the Board of Directors. Maintaining a portfolio based on one commodity could have negative implications for institutional investors. Diversification into other precious metals as well as a mix of T-bills and high-technology stocks appears to be in order.

The *Six Geese-a-Laying* constitute a luxury that can no longer be afforded. It has long been felt that the production rate of one egg per goose per day is unacceptable in the face of increased competition from Pacific Rim nations. Three geese will be let go, and outsourcing of this function, including consideration of offshore sources, is under advisement.

The *Seven Swans-a-Swimming* is obviously a number chosen in better times. The function is primarily decorative. Mechanical swans are on order, and the current seven will be offered retraining to enhance their outplacement.

As you know, the *Eight Maids-a-Milking* concept has been under heavy scrutiny by the EEOC. A male/female balance in the workforce is being sought. The more militant Maids consider this a dead-end job with no upward mobility. Automation of the milking process may free the Maids entirely. They would be allowed to apply as openings occur in divisions where a-mending, a-mentoring, or a-mulching are needed.

Nine Ladies Dancing has always been an odd number. This function will be phased out as these individuals grow older and can no longer master the steps.

Ten Lords-a-Leaping is overkill. The high cost of Lords, plus the expense of international air travel, prompted the Compensation Committee to recommend replacing this group with ten out-of-work congressmen. While leaping ability may be somewhat sacrificed, the savings are significant because we expect an oversupply of unemployed congressman in the next few cycles.

Eleven Pipers Piping and *Twelve Drummers Drumming* is a simple case of the band getting too big. Substitution of a string quartet, a cutback on original music, and requiring musicians to provide their own uniforms will produce savings that will drop right down to the bottom line. In the long term, digital audio transmission can displace these people entirely.

Though incomplete, time-motion studies indicate that stretching deliveries over twelve days is inefficient. If we can use just-in-time supply strategies and drop-ship in one day, inventory costs will fall and service levels will improve.

Lastly, it is not beyond the realm of possibility that deeper cuts may be necessary to maintain NPM's preeminence in the global gift-distribution business. Management will do what it must to prevent further erosion of the profit picture, including scrutinizing our new "Snow White" acquisition to reevaluate the necessity of having seven dwarfs when, maybe, two would do. More on this in a subsequent memo following the Executive Committee's return from its January retreat in the Bahamas.

Adapted from an Internet posting, November 19, 1996.

GETTING A LEG UP ON CORPORATIONS

When Yuletide rolls around each year, America is usually blessed with some terrific news story about a boss-with-a-heart to enhance the jolliness of the season, giving us all a much-needed chance to wallow in something positive.

December of 1996 brought news that John Tu and David Sun, founders of Kingston Technology in California, had sold a majority stake in their computer-chip enterprise for one-and-a-half billion big ones. The cheerful part is that these suddenly wealthy entrepreneurs did not merely sack up the money and run—they put $100 million of it into bonuses and special benefit programs for their 523 employees.

"Fruitcakes," screeched their high-tech corporate peers—"What in the name of Ebenezer Scrooge are you two up to?" Nothing. An amazed media found that this was no publicity ploy, no tax dodge, no manipulative management move—it was, simply, sharing. Sun and Tu quietly explained that, see, these creative, hardworking, loyal people at Kingston are what allowed them as owners to prosper, so the contribution of these folks should be respected and rewarded. Odd. "We want to show you can run a company by putting people first and not treating them just as expenses." Odder yet. But nice.

Aaron Feuerstein became 1995's corporate Christmas story the hard way: His fabric factory burned down. He was at a party on December 11 celebrating his seventieth birthday when the call came, sending him rushing into the night to stand helplessly as Malden Mills, founded in 1906 by his grandfather, went up in flames. Thirty-one hundred good-paying jobs went up with it, and the entire community of Methuen, Massachusetts, north of Boston, suddenly faced a bleak Christmas and a dark future.

Feuerstein surprised everyone, however. Instead of using the fire as an excuse to ship his whole shebang to Mexico, or to cash out and retire, he vowed that night to rebuild on the spot. But here is what really flabbergasted everyone: He also vowed publicly that he would keep all 3,100 employees on payroll, with full pay and benefits, while the rebuilding took place. Wall Street and its pundits howled in derision: "What's gotten into this fool?" Feuerstein said simply but firmly, it's the "right thing to do, and there's a moral imperative to do it irrespective of the consequences." Then he did it.

True to his word, everyone got paid for December, January, and February, even though no product was moving. By March 1996 some employees were able to start coming back to work on makeshift production lines, and the rest were put into a training program Feuerstein set up in a section of Methuen Mall to help them acquire advanced skills while they awaited reconstruction of the plant. When the rebuilt, fully modernized Malden Mills opened just over a year later, production already had been restored to prefire levels, demand for the factory's fabric was setting records, and 95 percent of the employees were back on the job.

In an interesting aside, a management professor at the mucho-prestigious Wharton School of Business, having witnessed the rebirth of this industrial phoenix from his distant ivory tower, later spoke in wonderment of the owner's commitment to the people of Malden Mills, as though Feuerstein had conjured up a completely new, even exotic, management philosophy: "Employees can be seen as an ultimate competitive advantage," the professor gushed. "If you treat them well, they'll pay you back in really hard work later on." This is why some say that "Ph.D." pronounced backward is "Duh!"

Tu, Sun, and Feuerstein each expressed amazement at the media hullabaloo over their actions. "What?" Feuerstein asked. "For doing the decent thing?"

Bingo. Such acts of responsibility are so beyond the norm, so antithetical to accepted boardroom practice, that when they occur their very oddity makes them hot-'n'-happening media features. Indeed, immediately after both of these heartwarming Christmas tales burst into national news, business analysts rushed forward to dump cold water on any fantasy that this is the way things should be done or even *could* be done in today's complex corporate economy. Kingston Technology and Malden Mills, the critics scoffed, are "privately held" companies, not owned by Wall Street investors as nearly all major corporations are. Messrs. Tu, Sun, and Feuerstein could be whimsical with company funds, these scoffers lectured, only because they were not subject to the short-term profit pressures that the executives of publicly traded corporations face from their big investors and Wall Street analysts. The "corporate system," they explained, has no room for beneficence toward employees, communities, or the environment.

None other than Milton Friedman himself, the heavily decorated establishment economist and patron saint of Wall Street excess, gave academic credence to such a myopic ideology by asking and answering his own rhetorical question: "So the question is, do corporate executives, provided they stay within the law, have responsibilities in their business activities other than to make as much money for their stockholders as possible? And my answer to that is, no they do not."

So, my question is, if the corporate structure exists only for stockholders, as Dr. Friedman makes clear, and since 80 percent of us are not stockholders, why should the larger public be so permissive toward this particular business structure, much less allow these entities to dominate our lives? You do not have to be the brightest light on the block to figure out that if the "corporate system" makes Aaron Feuerstein, John Tu, and David Sun the anomalies (and, in some executive quarters, the pariahs) of American business, then it's time to fix the system. I know the wisdom of the oft-cited aphorism, "If it ain't broke, don't fix it," but the equally sagacious corollary to that is, "If it *is* broke, run get the toolbox."

Remember Admiral James Stockdale, Ross Perot's hapless and clueless running mate in 1992, who opened his part of the Gore-Quayle-Stockdale vice presidential debate that year by posing these two revealing questions: "Who am I? Why am I here?" So, I ask, what is this thing called the "corporation" and why is it here? It is time to put this basic question back into political play. The corporation has become a given in our culture, not unlike smog, Republicans, and Wayne Newton—it has been with us so long we assume it has always been and always will be.

Not so. Even here in Texas, where the corporate logo is almost as exalted today as the hallowed Lone Star flag, and where state politicians eagerly play "go fetch" for corporate sponsors, the ascendancy of big business is relatively recent. Ours is a state founded by what the genteel would call your lower social classes: dirt-poor tenant farmers from the South fleeing debt, mavericks looking for space to be more maverick, outlaws, rebels, all-around hellraisers, and more than a couple of psychopaths—not sorts to embrace authoritarian structures or trust amalgamations of money and power. Quite the opposite—in 1845, the founding constitution of the State of Texas specifically *prohibited* the creation of banks! It also required anyone wanting to form a corporation to trudge up to the legislature and get two-thirds of both houses to approve the entity—a hurdle that was rarely cleared.

Nor was Texas especially quirky on this. It is not mentioned in the standard textbooks of American history, but practically all of our nation's Founders were appropriately anticorporate, and at the time of the Continental Congress, only about forty of these suspect corporate critters had been allowed to form in our land, and those were kept on a very short leash.

Like powdered wigs and boiled beef, the corporation is a British invention, essentially created by the Crown as a legal vessel to amass the capital needed to exploit the wealth of its American and Canadian colonies. Operating on the principle that it takes a dime to make a buck, the corporation—or "joint stock company," as it was then known—was a paper structure devised to collect the dimes from English investors, organize the looting in the colonies, and return the dollars to said investors.

Of course, looting was nothing new, but this "joint stock" connivance was a devilishly radical social departure. For the first time, *ownership* of an enterprise was separated from *responsibility* for the enterprise. If an individual business person loots, pollutes, or otherwise behaves illegally, he or she is individually accountable to the community for those actions—that is, you can have your sorry ass hauled into court, be fined, be put out of business, be tossed in the slammer, or all of the above. But the corporation is a legal fiction that lets the investors who own the business skate whenever the business behaves badly (read: steals, kills, poisons, pillages, corrupts, and so on), avoiding individual responsibility for illegal actions done in their name, even when such actions profit them mightily. Not only can these aloof owners have their cake and eat it, too, they can also eat yours and laugh all the way to the bank.

Like letting a cat loose in a fish market, the corporate structure invites mischief, which is why the powdered wigs on our side of the Atlantic were so stingy about issuing charters for them and so specific about what actions the charters authorized. As Richard Grossman, author of *Revoking the Corporation,* and Jonathan Rowe, contributing editor of the *Washington Monthly*, have reported in their investigative writings, the Founders believed that corporate charters should be authorized only to serve "the public interest and necessity." Through most of the nineteenth century, state charters typically limited the corporation to one kind of business, prohibited it from owning other businesses, strictly limited the amount of capital it could amass, required the stockholders to be local residents, often vested part ownership in the state itself, spelled out specific benefits the corporation had to deliver to the community and even to particular groups of people (farmers and fishermen, for example) and put a twenty- to fifty-year limit on the life of the charter. And, imagine this, legislatures were not shy about yanking charters when a corporation went astray from its mission or acted irresponsibly.

What a sensible system! So what went wrong?

As an old "*dicho*"—a Mexican maxim—puts it: "*Poderoso cabellero es Don Dinero.*" Money is a powerful gentleman. Just as America's public leaders clearly saw the dangers of the corporate entity, the avaricious saw its phenomenal potential to help them make a killing, and they began early on to lather up politicians with money to try to loosen up the chartering process. (Fast-forward to 1985: A Texas state senator not above using his position to line his own pockets once said to me in an inebriated moment of candor, "I seen my chances, Hightower, and I took 'em." That is about as philosophical as the corporate power grab gets.)

During the Civil War, numerous corporations were chartered to supply the Union Army, and the commander-in-chief, Abraham Lincoln, did not find it a positive experience. In an ominous foreboding of corrupt practices by today's Pentagon contractors, many of these corporations delivered shoddily made shoes, malfunctioning guns, and rotten meat. Honest Abe viewed the rise of corporations as a disaster, penning these thoughts in an 1864 letter:

> *I see in the near future a crisis approaching that unnerves me and causes me to tremble for the safety of my country. As a result of the war, corporations have been enthroned and an era of corruption in high places will follow, and the money power of the country will endeavor to prolong its reign . . . until all wealth is aggregated in a few hands, and the Republic is*

destroyed. I feel at this moment more anxiety for the safety of my country than ever before, even in the midst of the war.

Sure enough, during the next three decades, assorted industrialists and corporate flimflam artists known collectively as the robber barons were enthroned, taking hold of both the economy and the government. Corruption did abound, from state houses to the White House (so much so in President Ulysses Grant's term that he was compelled to issue a public apology to the people for the tawdriness of his administration), and the concentration of wealth in the grasping hands of such families as the Astors, the Vanderbilts, the Rockefellers, and the Morgans reached proportions unheard of . . . until today.

Two major changes unleashed the modern corporation on us. First was the emasculation of the state corporate charter, removing bit by bit and state by state all restrictions imposed to protect the public. In 1890, for example, a lawyer named James Dill enticed New Jersey's governor to offer the most permissive charter then available from any state. Dill's enticement was both political and personal: (1) If the governor went along, corporations would come running to Jersey and gladly pay large chartering fees that would swell the state's treasury and make the governor look like a fiscal genius; and (2) Dill would create a company to process the charter applications, and the good governor would get a piece of this lucrative action. The governor seen his chances, and he took 'em.

Thus began a race among the states to see who could give away the most to lure corporations, which were now multiplying like fleas. The race to the bottom ultimately was won by Delaware, a DuPont fiefdom that essentially lets corporations write their own charters and provides corporate protections *from* the public, rather than for the public. In the April 1996 *Washington Monthly*, Jonathan Rowe described the result: "By the time of the Great Depression, Delaware had become home to more than one-third of the industrial corporations on the New York Stock Exchange; 12,000 corporations claimed legal residence in a single office in downtown Wilmington. . . . By the mid–1970s, half the nation's largest 500 corporations were chartered in tiny Delaware." The total physical presence that most of these "Delaware" firms have in the state is a file folder.

Today the corporate chartering process is so perfunctory that it can be handled by a phone call. In the February 1997 issue of *American Airlines* magazine, a company that processes corporate charters ran an ad next to one promoting "Tattoos with Your Logo" and across from one for "Hair Transplantation" (call 800-NEW-HAIR). The chartering outfit announced in bold type: "You Can Form Your Own Corporation by Phone, in Any State." "It's amazing but true," the ad copy avers. "You know the advantages of incorporating: Incredible tax breaks. Protection of personal assets. . . . You can incorporate in 8 minutes . . . over the phone . . . FOR AS LITTLE AS $45. . . . We can incorporate you in any state and are most famous for our ability to set you up as a Delaware corporation—the well-known corporate haven."

The other big change came in 1886, when the U.S. Supreme Court essentially made the corporation bulletproof. In a stunning and totally irrational decision made without bothering to hear any formal arguments (indeed, the learned justices said they *did not want to hear* any arguments—case closed, minds shut, go away), the Court abruptly decreed in a case brought by a railroad company that a corporation is "a person," with the same constitutional protections that you and I have, including the right to free speech, which in turn has been interpreted as the right of the wealthy, the powerful, and the corporate to buy politicians.

Dr. Frankenstein could not have done better than the courts and legislators. In only a century, the corporation was transformed into a superhuman creature of the law, superior to you and me, since it has civil rights with no civil responsibilities; it is legally obligated to be selfish; it cannot be thrown into jail; it can deduct from its tax bill any fines it gets for wrongdoings; and it can live forever.

There is a useful saying handed down from the old mountain men: "Never drop your gun to hug a grizzly." That's the problem—our politicians are hugging the grizzly. In the spring of 1996, with corporate executives rapaciously whacking jobs and abandoning communities, with Pat Buchanan stinging the political establishment with his rousing call to the anticorporate barricades, with even the gray-flanneled *New York Times* running a six-part, front-page series tut-tutting about the economic and social damage being wreaked by corporate excess—the party of Teddy Roosevelt and the party of Franklin Roosevelt both sipped cocktails in the bear's den, way too involved in wheedling for campaign funds to give a moment's thought to taking any action to tame the beast.

The two parties could not even bring themselves to debate the issue of corporate rapaciousness in the presidential campaign. The ossified Republican, Bob Dole, mostly expressed befuddlement: "I didn't realize that jobs and trade and what makes America work would become a big issue," the Dolester mumbled just before the New Hampshire primary, which he lost. Bill Clinton, on the other hand, the standard-bearer of the once-proud party of the working class, said that he could feel the people's pain and that, by golly, he had just the fix for them: He would convene a White House "Conference on Corporate Citizenship."

He did. Held on May 16, 1996, the day-long event was no more than a dog-and-pony show held so the media would cover about a hundred corporate slicks telling the Prez that, by golly, they felt the people's pain, too. Tough turkey, though. Dwane Baumgardner, CEO of Donnelly Corporation, did not mention that on the very day he was sitting there with Clinton and the other conferees clucking their tongues, his company back in Tennessee was announcing it was shutting down a plant and punting about a hundred workers. The only tangible result of Clinton's conference was the creation of an annual "Ron Brown Corporate Citizenship Award," named for the former corporate lobbyist and Secretary of Commerce whose tenure was marked by an energetic effort to further globalize and eliminate American jobs. And that was that. The President hugged the grizzly, and there was no more talk of corporate irresponsibility during the campaign.

"Corporations have become the dominant institution of our time, occupying the position of the church of the Middle Ages and the nation-state of the past two centuries." This is not the wail of Ralph Nader or some other corporate critic, nor is it even a criticism; it is the wholly approving observation of a new, up-tempo business magazine called *Fast Company*.

What are the world's biggest economies? The United States is at the pinnacle, of course, and Japan, Germany, the United Kingdom, and China are top-ranked as you would expect—but Mitsubishi? It is number twenty-two, ranked ahead of Indonesia. General Motors is number twenty-six, bigger than Denmark or Thailand. Ford is thirty-first, ahead of South Africa and Saudi Arabia. Toyota, Shell, Exxon, Wal-Mart, Hitachi, and AT&T each are in the top fifty. Indeed, Sarah Anderson and John Cavanagh of the Institute for Policy Studies have found that of the one hundred largest economies in the world, fifty-one are corporations. Further, the economic growth of corporations is exceeding that of nations, and their reach knows no bounds.

If you read the business pages, you will find the captains of industry unabashedly exulting in their enthronement, declaring their independence from any social contract or community mores, and from nations themselves, even the most powerful nations: "National Cash Register is not a U.S. corporation," CEO Charles Exley has announced, "it is a world corporation that happens to be headquartered in the United States." Likewise, Unocal Corporation says it is de-Americanizing by opening a "twin corporate headquarters" in Malaysia, and other U.S. firms now refer to themselves as "multidomestic corporations." A shipping executive recently expressed the prevailing, supercilious attitude of today's corporate peerage when he proclaimed: "It's a new world. All bets and all past practices are off."

Ralph Nader told me about a tense and telling moment at the annual stockholders' meeting of General Motors in 1996, held in Wilmington, Delaware (where else?). The setting was appropriately auspicious, with the board of directors arrayed like ministers of state across an elevated platform in a grand ballroom, the GM logo looming above the proceedings, CEO John Smith in imperious pose at center stage, and row upon row of stockholders splayed across the floor below. In the midst of this pomp and ceremony, a lone stockholder gained recognition to speak and pose a question. Noting that there was an American flag on display to the side of the platform, and observing that GM has eliminated some 73,000 U.S. jobs while creating almost the exact same number in low-wage countries in the past decade, the stockholder asked politely if CEO Smith and all members of the board would rise and join him in the simple gesture of pledging allegiance to the flag.

There was some embarrassed tittering among the board members, a scurrying of legal counsels back and forth behind the podium, a bit of hemming and hawing by the CEO, but the bottom line was no, they would not. Reporters were there with their TV cameras and their computer PowerBooks recording the moment, but not one whisper of this revealing exchange made the news.

No single step, no magic fix-it, is going to bring the arrogance of these elite investors and managers back to earth, but one strong hammer our democracy needs to have in its toolbox is a corporate charter with teeth. Just as the larger community specifies what it expects of poor welfare mothers, so must we again set strict terms for the much more generous privileges bestowed on these haughty and domineering creatures of the state, including making corporate board members and managers individually responsible for the malfeasance of their enterprises and reestablishing term limits on each corporation's charter. Such a meaningful public charter, along with a straightforward six-word constitutional amendment that says, "A corporation is not a person," would begin to bring the destructive, single-minded profit drive of the corporation back into balance with the greater goals of our society.

This ain't gonna be easy, to say the least. As a Republican state senator from Waco, David Sibley, said about a complex bill he was trying to maneuver through the Texas legislature a couple of years ago, "This is harder than playing pick-up sticks with your butt cheeks." (Now we know what Republicans do behind those closed doors marked Members Only!)

Challenging hegemonic corporate power head-on is made all the more difficult by the combination of gutlessness and whorishness that prevails among our present political leaders. This is a struggle that has to be organized and fought with no expectation of help from those in high positions—not only are the odds against us, but some of the evens are, too.

On the bright side, people already are focusing on the target, organizing, and fighting back, with newly aggressive groups like ACORN; AFL-CIO; the Alliance for Democracy; the Institute for Local Self-Reliance; the New Party; and the Program on Corporations, Law and Democracy in the forefront.

And when we think it is too difficult, even impossible, to win such a battle, it is worth remembering that Americans just like us were here before . . . and have won. Remember that everything

Lincoln foresaw—enthronement, corruption, aggregation, and destruction—came true over the next forty years, except the last: destruction of the republic. This is because ordinary people rose up against the corporate giants, the plundering families, and the corrupt politicians. In heroic and historic battles from the early 1880s into the early 1900s, the republic was saved from the overreaching and overbearing robber barons by the likes of Mary Ellen Lease, called Mary "Yellin" Lease by the establishment that feared her. She was popularly known as the "Kansas Pythoness" for the kind of oratorical sting she displayed at the founding of the People's Party in Topeka in 1890:

> Wall Street owns the country. It is no longer a government of the people, by the people and for the people, but a government of Wall Street, by Wall Street and for Wall Street. . . . Our laws are the output of a system which clothes rascals in robes and honesty in rags. . . . The people are at bay, let the bloodhounds of money who have dogged us thus far beware.

Her war cry to rally oppressed farmers on the Plains was: "It's time to raise less corn and more hell."

She was joined by such leaders as James B. Weaver, "Sockless Jerry" Simpson, Ignatius Donnelly, and Charles Macune in the Populist Movement; Samuel Gompers, Big Bill Heywood, Mother Mary Jones, Joe Hill, Eugene Debs, Clara Lemlich, Elizabeth Gurley Flynn, and many, many more in the labor movement; W. E. B. Du Bois among African-American leaders; Upton Sinclair, Ida Tarbell, and Lincoln Steffens in the muckraking press; and thousands more national, regional, state, and local leaders.

Together they forged a new politics that elected populists, socialists, radicals, and other noncorporatists to legislative seats, governorships, and Congress, and while they never elected a President, they did force both the Democratic and the Republican parties to embrace their reform agenda, producing the nation's first trust-busting laws, the first wage-and-hour laws, women's suffrage, the first national conservation program, the direct election of U.S. senators, and other populist, working people's reforms.

What sparked the success of this grassroots movement was not merely widespread recognition that things were bad, that the many were being had by the few, but the realization that *it does not have to be this way,* that people themselves can take charge.

Yes, corporations today are conglomeratized, globalized, and more powerful than any robber baron could have dreamed, but they are not part of the natural order—they remain an artifice of government. The Supreme Court also ruled in 1906 that "The corporation is a creature of the state. It is presumed to be incorporated for the benefit of the public." When it ceases to be a benefit—declaring itself above the common good and beyond the common decencies of John Tu, David Sun, and Aaron Feuerstein—then we *can* cease to sanction its incorporation.

As powerful as the corporation seems, remember this: No building is too tall for even the smallest dog to lift its leg on.

VERNON JORDAN'S DREAM

I was there the day Vernon Jordan gave his "I Have a Dream" speech.

By the time he finished, there was not a dry seat in the room.

Jordan is a hell of a guy—smarter than a tree full of owls, he always has a welcoming smile to greet you, and he can pound a podium with the best of them. He had been a savvy and important civil rights advocate in the 1960s, eventually becoming head of the Urban League, which is where he first rubbed up against the serious money of Fortune 500 companies, several of which were helping to finance the league. Vernon discovered that he looked good in green, and the companies clearly were impressed with him, so by the 1980s he had made the leap from the streets to the suites, becoming a savvy and important *corporate* advocate.

Ever since, he has been comfortably ensconced in the high-powered lobbying firm of Washington political doyen Bob Strauss, and he quickly became a major player among Washington and Wall Street insiders, helping take care of business for such clients as Philip Morris, American Airlines, and dozens more. Next thing you know, he finds himself a full-fledged member of the corporate class, not only lawyering for the bigs, but also sitting on the boards of directors of . . . drum roll please:

American Express
Bankers Trust
Corning
Dow Jones
JCPenney
Revlon
RJR Nabisco
Ryder Systems
Sara Lee
Union Carbide
Xerox

As Jordan made his way up in the corporate world, he crossed paths with another comer, Arkansas governor Bill Clinton. The two hit it off, and Jordan later became famous, then infamous, as First Pal of the President. Besides sharing the altruistic goal of helping White House interns like Monica Lewinsky get jobs, Jordan and Clinton also share Southern roots, a passion for golf, and, some say, a near-obsessive interest in sex. According to a *Newsweek* report, when Jordan was asked what he and his buddy Bill discuss on their frequent golf outings, he said: "We talk pu—y"—and I don't think *Newsweek* meant puppy. Whatever, he has long served as trusted family confidant to both Bill and Hillary, as a behind-the-scenes adviser with unparalleled clout in the Oval Office, and as a combination of *consigliere*, financial rainmaker, and political cheerleader for Clinton.

It was this last role that brought Jordan to New York City in July 1992, where he gave the speech that still burns in my memory. He was in the city to help orchestrate the cheering for Clinton's nomination to the presidency at that year's Democratic National Convention.

Enter from backstage: me. I was a Texas delegate to the convention—actually, a "superdelegate," a technical designation that meant nothing except that my fellow Texans kept referring to me with a derisive snicker as "Your Superness." Having been to three national conventions and been privileged to speak at two of them, I was looking forward to this one because I had no heavy duties and could mostly just kick back and enjoy it. In the Democratic primaries, I had supported Senator Tom Harkin's failed bid for the nomination, then moved to Jerry Brown when Harkin bailed out, so I was hardly on Clinton's A-list in New York. Indeed, I was one of the few delegates who did not claim to be an F.O.B. (Friend of Bill's), if not actually claiming to have been his roommate at Oxford (of *course* he didn't inhale, the room was too filled with roommates to breathe!).

I was such a latecomer to Clinton's bandwagon largely because I did know him politically. As Texas agriculture commissioner during the family-farm crisis and tractorcades of the eighties, I had visited with Governor Clinton in Little Rock and tried to get him to take up leadership of the farmers' cause nationally. No go. Just as he now is from the White House, I found Governor Bill eager to be "for the small farmers of Arkansas"—as long as this did not require him to do anything that might actually help them, such as taking on the bankers who were choking the life out of them or challenging the big processors that were cheating them. When it came time to deliver, the guy was all hat, no cattle.

Still, as we used to say in high school, you don't have to be in love to dance. By convention time, Superdelegate Hightower of Texas was ready to dance with the Arkansas Traveler or anyone else who could deliver the White House from the clutches of George Herbert Walker Bush II. While I knew that Clinton was no FDR, no LBJ, and maybe not even a Jimmy Carter, I felt that at least he would not push Republican policies to crush the middle class and the poor. (OK, so I underestimated the big galoot, but it seemed a fair assumption at the time.)

As soon as I arrived in New York, though, I knew that something major had happened to my party. I opened my delegate's packet and right there on the official program of the 1992 Democratic National Convention was this startling proclamation:

HOSTED BY
AT&T, AMERICAN EXPRESS,
NYNEX CORP. AND TIME WARNER INC.

Great galloping ghosts of Thomas Jefferson, Andy Jackson, and William Jennings Bryan—was I at the Republican convention? Now, I had been to the state fair twice and had been a Texas politician for a decade, so I did not come to New York a virgin. I was fully aware that Wall Street's finest have long had a backstreet affair with the Democratic hierarchy, but Clinton's convention took the relationship to a different level altogether, formally and publicly wedding the "party of the people" to the bigamist corporate powers that had long ago tied the knot with the GOP.

This wedding was, of course, about money and access. The four "principal sponsors" had put up $400,000 each to headline the '92 convention. Dozens of others—including Philip Morris, Shell Oil, Coca-Cola, Archer-Daniels-Midland, and ARCO—had shelled out a minimum of $100,000 each to become "managing trustees" of the party.

Trustees? Since when do Democrats have trustees? I asked

around, and none of the elected delegates I talked to knew there were such privileged positions, much less that they were for sale to the very same corporate elites that are kicking the stuffing out of working families and others who form the core of the party's delegates and voters. More infuriating, while Democratic delegates were lucky to catch a glimpse of Clinton or Gore as the daring duo dashed about town in their limousine convoys, the trustees were getting major face time with the two big guys all week, talking policy over eggs Benedict at small, very private brunches, or just schmoozing one-on-one over late-night cocktails in the President-to-be's suite. It was a foreshadowing of the infamous White House coffee klatches and Lincoln Bedroom "overnights" that Clinton would hustle up for big corporate donors in '95 and '96.

Technically, the convention was held in Madison Square Garden, but there seems to have been a preconvention that took place inside the New York Stock Exchange, with major corporations buying shares in the party as fast and furiously as the conventioneers were now buying trinkets and T-shirts to take back to Tallahassee, Keokuk, and Tacoma. Robert Rubin, who would become a top dog in Clinton's cabinet but was then a top dog at Goldman Sachs (Wall Street's leading marketer of government bonds), chaired the Host Committee and had put out a "*Buy!*" signal on the convention. There even was an "adopt-a-delegate" program that allowed companies to purchase the right to sponsor entire state delegations—we Texans were the property of Barnes & Noble, for example, Florida belonged to Merrill Lynch, Michigan was picked up by Met Life, both North Dakota and West Virginia were acquired by Pfizer (take two, they're small), and on down the list. In addition, every coffee, breakfast, brunch, lunch, high tea, happy hour, cocktail party, and any other event involving edibles or potables had its own corporate sponsor, with several of the company's slicks-in-suits standing around with forced smiles trying not to get their shines messed up by us riffraff who were merely delegates.

These were the *public* gatherings. The printed schedule of the week's events also was heavily larded with "by invitation only" functions, including the R. J. Reynolds brunch at the St. Regis for Senator Wendell Ford, the "Baby Bell" phone companies' reception for then-speaker Tom Foley at Cafe Iguana, the Kerr-McGee breakfast at the Plaza honoring Senator David Boren, the Goldman Sachs soirée (there is that bond connection again) at the Museum of Modern Art paying tribute to Democratic governors, the Natural Gas Association's dinner-dance cruise celebrating the then-chair of the Senate energy committee aboard the yacht *Princess,* Paine-Webber's romp at Tavern on the Green for Democratic senators, and—my personal favorite—a gaggle of corporate sponsors hosting the Senate committee chairs at, appropriately enough, the Central Park Zoo.

With dozens more of these exclusive events held at New York's most fashionable watering holes, nomination week was transformed into a corporate fund-raiser interrupted only briefly by a convention. We practicing Democrats, who were kept on the far side of the velvet ropes at these private affairs, were not even allowed to peer into the goings-on inside, but we could hear plenty of big, wet smooooching sounds. Just the sheer volume of corporate functions and functionaries had to convince even the most gullible of us delegates that the party might be over for us. As blues master Albert Collins put it in one of his songs: "Too many dirty dishes in the sink for just us two/You got me worried, baby/Who made them dirty dishes with you?"

Then there was the train. Any delegates who felt at all squeamish about the garish corporate takeover of the convention upchucked on their shoes when they heard about Victory Train '92. Congressional and party bigwigs rode to New York from Washington's Union Station aboard a specially chartered train . . . paid for by corporations and lobbyists. "Once aboard, you'll be able to roam the train and enjoy the ride with Members of Congress, Democratic governors and mayors . . ." promised a promotional piece from party headquarters, selling space to lobbyists for $25,000 and up.

The ride was closed to the media, but Sheila Kaplan of *Legal Times* managed to scramble on board and filed this report: "Although the DNC would not divulge the names of its big-ticket donors, a stroll through the train revealed that [lobbyists] for many of the nation's biggest companies had taken the DNC's bait. Among them: Stephen Paradise, senior VP for congressional and regulatory relations of the New York Stock Exchange; Paul Equale, senior VP of government affairs for the Independent Insurance Agents of

America; Sharon Spigelmyer, manager of federal services for Arthur Andersen and Co.; and Harry Wiles II, VP of federal affairs for the Wine and Spirits Wholesalers of America."

Pentagon contractors, the Bell phone companies, banking interests, and others took the joyride, too, imbibing freely from the antique bar car, snacking from silver platters of munchies, and, mostly, taking advantage of four uninterrupted hours of access to top lawmakers and other public officials from whom they needed favors. As one lobbyist told Ms. Kaplan, "What was the most fun was that they were a very captive audience."

On Wednesday morning of convention week I was to address the breakfast meeting of the Wisconsin delegation, bivouacked at the St. Moritz Hotel. Miller beer was the morning's sponsor, in charge of bringing home the bacon to the Wisconsinites, though many of the hale and hearty delegates of this fun-loving state commiserated with me that the brewer's own product was not on the table—breakfast of champions.

Three of us were set to speak that morning and, as the time drew nigh, Vernon Jordan leaned into me and asked if I minded if he went first, for he had many fish to fry that day. Not at all, I responded, and am I glad I did, for it gave me a front-row seat for Vernon's "I Have a Dream" revelation.

He lit up the room with his personality and passion, going on for some fifteen, twenty minutes about the virtues of his personal friend, the "Man from Hope." I was as mesmerized by Vernon's eloquence as the rest of the room, applauding as he reached full oratorical throttle on the do-or-die necessity of putting Clinton in the White House. Then, in a sudden, theatrical move, Jordan reached across the podium and pointed. Was it a teacher he was singling out, a factory worker, a senior citizen, a child?

No, he was pointing to one of the Miller beer placards that our corporation-of-the-moment had placed on each table. "This," Vernon pleaded, "This"—and someone hesitantly lifted a Miller placard to this powerful man's outstretched hand. "This," he said, finally clutching it, "is why we must have Bill Clinton, a Democrat, in the White House. Miller beer is owned by Philip Morris, and I represent them in Washington. I want you to know that I'm tired of taking my clients in to see a Republican

President, I want them to be able to walk into the Oval Office and be greeted by a Democrat!"

And you say there is no idealism, no crusading spirit, no dreamers left in the Democratic Party. Vernon Jordan has been to the mountaintop—and I was there when he came down to tell us about it.

I still shiver at the memory.

HIGH-TECH EXCESS

When I was growing up, my folks had a magazine-wholesaling business, which meant that twice weekly they hauled heavy bundles of every publication from *TV Guide* to *Playboy* to the newsstands, grocery stores, and other retail outlets throughout the greater metropolitan environs of Denison, Texas. When I was about thirteen, once I finally got enough heft on me to outweigh a bundle, it became one of my family duties to help load Daddy's panel trucks for the next morning's delivery. I was thrilled, because the job also required me to back the trucks down the alley and into the loading area—my first experience behind the wheel of a moving vehicle.

The fact that I learned to drive backing up might explain my failure today to appreciate fully the plethora of blessings showered upon us by the "technological revolution," the "digital age," the "information superhighway," and the "bridge to the twenty-first century."

Not that I am antitech. No, not at all. True, I have no VCR, CD player, or even a doorbell, and I am absolutely hostile both to call-waiting and to the use of cell phones in public places (a death-penalty offense, in my opinion). Sitting on the Internet for hours holds about as much appeal for me as sitting on the Interstate. However, twist-off beer caps—now *that's* progress. Also, the fax machine—I worship it. I have an answering machine, too (and wish everyone did), a mulch mower, an automatic defroster on the rear window of my Ford and—check this out—my own Web site (http://www.jimhightower.com)!

But let's back up a minute. High tech is getting way out of hand, put forward as "the answer" by a lot of people who do not seem at all clear on the question. We have bad television in this country, for example, so Wall Street and Washington have worked out a technological plan to deliver 500 channels of it on new, high-resolution TV screens. Hello?

As Hemingway wrote, "Never mistake motion for action." Likewise—despite what we are told by corporate hustlers, presidents, assorted "futurists," Ph.D.'s, Newts, *Good Morning America,* psychics, Bill Gates (see number one), and other carny barkers of the New Wave—never mistake technological advances for progress. I offer the following three points about parking meters, penises, and peppers in evidence, and rest my case.

PARKING METERS

Herewith a story of villainy and heroism, starting with the first. Mr. Vincent Yost of Philadelphia undoubtedly would disagree that he is a villain, probably even counting himself in the company of Benjamin Franklin, Eli Whitney, and Thomas Edison as technological tinkerers who have improved the human condition. But you be the judge. If you thought nuisance technology surely had bottomed out with the advent of car alarms and automated phone solicitations, you had not figured on Mr. Yost's high-tech parking meter.

I start from the position that the parking meter is not our friend. It tends to expire most inconveniently, usually resulting in a fine, but even when we operate within the time limit imposed by the damn thing, it still amounts to taxing us to park at curbs we have already been taxed to build.

But existing parking meter technology does occasionally allow us one of the few, truly triumphant moments humans get in urban life: finding an empty space with time left on the meter. When I pull in and find ten minutes or so gratis, I cannot help but pump my fist and shout, "Yes! This is my day."

It is this small joy that Vincent Yost wants to take from us. His company, Intelligent Devices Inc., is testing and preparing to market a new line of fiendish, computerized meters that will use infrared sensors to erase any time that is left on the meter when a car pulls out of its space. Worse, Yost's impudent contraption will not let you hold on to your parking spot by adding more money to the machine. Oh yes, you can feed quarters into its maw to your heart's content, but nary a minute's credit will it register, while it glares defiantly at you with its single electronic eye.

This "intelligent meter" (Yost's term) also keeps track of how long your car has been parked at an expired meter. "No more running up [to a ticketing officer] and saying 'It just ran out,'" brags this pitiless technician, diabolically chortling, "The meter doesn't lie." As an added twist of the screw, he makes it easier for the ticketers to nail you by replacing the subdued red flag of today's devices with a bright red light.

No way, you say? Look for it at a parking space near you soon. Cities make a ton on our parking meter change and fines, and the prospect of collecting more is like a vision of sugar plums dancing in the heads of local officials. Yost has the numbers to excite them—a test of his smart machine in Lower Merion Township, Pennsylvania, jacked up the average weekly take per meter from $12.45 to $44. This year he will position 200 of what he impishly calls "ruthlessly efficient" instruments around the country as demonstration projects requested by a half-dozen cash-hungry city bureaucracies, from New York City to Kansas City.

Turn now to the heroine of our story, who is a criminal. At least that is the view of the gendarmes of Cincinnati. Of course, many of America's folk heroes have been outlaws—Jesse James and Pretty Boy Floyd come to mind—and now we can add the name of "Ma" Stayton, the Meter-Feeding Granny.

A sixty-three-year-old grandmother, Sylvia Stayton was headed into a Kinko's store in Cincinnati's Clifton neighborhood last year when she noticed that a policeman was poised to write a ticket. Before he could do the deed, however, Ms. Stayton slipped alongside him, asked that he spare the poor soul, and deposited a nickel in the red-flagged device. Noticing that the next meter had expired, too, she stepped over and put a dime in it. Fifteen cents' worth of kindness. She said later, "I hope somebody would do that for me."

The officer was not charmed. He told Ma Stayton that it was a civil violation in Cincinnati to put money in expired meters, and he asked for her ID. She did not give it, thinking he was kidding. He was not. He told her she was under arrest. Thinking he had now gone from kidding to silly, she started to walk away. Bad move. He grabbed her, twisted her arms behind her back, *handcuffed* her . . . and hauled Granny to the pokey, where she was booked for "interfering with the duties of an officer" and tossed in a cell.

Nor did cooler heads prevail higher up in Cincinnati's hierarchy: "We're not dropping the case because she's a grandmother," snapped the city prosecutor, who promised to bring down the hammer of the law on this promiscuous meter-feeder. The authorities tried to get her to accept a plea bargain, in which she would admit to guilt but receive no fine. Stayton stood her ground, though, saying simply: "I didn't do anything wrong." She instantly became a champion to all of us fed up with being pushed around by machines and the technocratic authorities who tend them. A "Free Sylvia" movement swept the city, complete with a T-shirt picturing her behind bars made of parking meters and the line, "Guilty of Kindness," emblazoned underneath. A "Sylvia Stayton Legal Abuse Fund" was started, and the *Gary Burbank Show* on radio station WLW even came up with the ballad of the "Grandma Meter Maid," sung to the Marty Robbins classic "El Paso":

The Grandma was adding more time on the meter.
The policeman said "Lady, you're breaking the law!"
But Sylvia ignored him and dropped in a quarter,
Down went his hand for the cuffs that he wore. . . .

It's all great fun, unless you are Sylvia. She was forced to hire an attorney and face trial on February 5 of 1997. The city put two prosecutors on the case, and they went at her like she was Ma Barker the bank robber, not Ma Stayton the meter-feeder. So determined were they to make an example of her that they spared no expense in amassing a legal case that took two days to present to the jury. Of course, they only succeeded in making an example of themselves and Cincinnati (Town Motto: "We go the extra mile to crush kindness and look really, really, extraordinarily, supercalifragilisticexpialidouciously STUPID"). Sure enough, Sylvia was declared guilty of interfering with an officer by putting 15 cents in two parking meters, and the judge slapped her with a $500 fine, which was more than covered by outraged citizens across the country who heard her story and sent checks, along with their heartfelt thanks for standing up for us humans against autocratic parking meter codes.

Sylvia Stayton's name will live on as a heroine of the human struggle, but the technocracy is relentless: Cincinnati officials have ordered a test installation of Vincent Yost's parking meters.

PENISES

It seems like only yesterday that technologists of the fashion, fragrance, and body-enhancing industries pretty much left us men alone, focusing their dollar-sign eyes solely on the makeover of women.

Now, though, the same fashionists who developed the Wonderbra are pushing their latest breakthrough, the Super Shaper Brief. Essentially, these are padded jockeys to enhance the assets of the male bod—undies with sewn-in, seamless posterior pads to make us guy-types appear to have firmer, rounder, tighter buns, for $24.95 a pair. And for you men who worry that you were shorted by nature, for five bucks extra, Super Briefs offer an optional snap-in codpiece, or "endowment pad," as they prefer to label it.

Naturally, men stink. But a little deodorant and aftershave won't cut it anymore. The fragrance industry insists on enhancing the male smell, so olafactorists (oh yes, there is such a branch of industrial technologists now) are constantly scouring the globe for new aromatic ingredients to produce tomorrow's scented sensation. Calvin Klein's Escape for Men has been such a big seller, say the scientists who concocted the stuff, because it smells like "birch twigs sweating in a sauna." I guess that's better than the smell of "Bubba sweating in a sauna," but I'm not sure. The big fragrance news, however, is the recent discovery of an ingredient taken from a rare Costa Rican insect that will let us "smell just like ambergris." Great! Can't wait. What is ambergris? I looked it up: "an opaque secretion of the sperm whale."

At least those enhancements do not hurt, which is more than can be said for phalloplasty. The vanity industry that brought our culture the horrors of silicone implants and other breast enlargement surgeries has now turned its attention to an essential national need: "penile enlargement." Apparently the lengthening operation was first developed by a Chinese doctor with the almost-eponymous name of Long Daochao, and was brought to the United States in the early 1990s, where it nicely complemented a penis-widening process developed in the late eighties by a Miami doctor whose previous specialty was silicone lip enlargements.

Do we need this? What happened to the true-life wisdom of the old blues song, "It ain't the meat, it's the motion"? Never mind logic, though; apparently some men are willing to believe that artificial augmentation will improve their lives—the *Wall Street Journal* reports that phalloplasty has become a major profit center for several plastic surgeons, former hair-transplant entrepreneurs, some urologists, and . . . well, who knows?

Practitioners of penile enlargement do not seem to follow standard surgical protocol, instead openly soliciting with ads in local sports pages and in such men's magazines as *Penthouse,* asking customers to call phone numbers like 1-800-BIG-MALE. The procedure is not taught in any university, is mostly performed in private offices rather than hospitals, is not monitored by any federal or state agency, and is so risky that malpractice insurers rarely cover physicians who perform it.

As you might expect, this "medical specialty" presently finds itself awash in malpractice suits by patients whose complaints range from disappointing results to impotency. Still, there are some customers who claim to be happily enhanced, like Frank Whitehead, a Southern California entrepreneur who went in for both a lengthening and a widening job. He told *Journal* reporter Lisa Bannon that the makeover really has put a strut in his step: "Now [when] I go into business meetings, I'm thinking, 'If you guys had just half of what I have.'" (It is said that when someone loses sight or hearing, other senses increase to compensate for the loss, but I wonder if the converse is true, if penile enlargement, for example, necessarily produces a commensurate shrinkage in IQ? Someone should study this.)

Meanwhile, there are men suffering from a very serious condition called "erectile dysfunction," resulting in a debilitating and often heartbreaking impotency. Once more, though, technology looms to the rescue with a prescription drug that apparently produces its desired effect on the male organ in 80 percent of all cases. Good for science! This uplifting news, however, is tempered by certain practical shortcomings, such as the fact that the drug is not taken orally, but in the form of a shot.

"No sweat," you say, "I can take a shot." Sure you can, Big Guy, but this is not a shot in the arm or even in the tush. You take this drug by injecting yourself in the very place that probably would be tops on your list of places you would least like to have injected: your very own personal sexual organ.

Imagine the scene: romantic music, lights low, a little wine, she is in the mood, the passion is palpable, the moment is nigh, and you say, "Excuse me, I'll be right back." The next voice she hears from the bathroom is yours: "Eeeeeeeeeyowwwwwwwwwww!"

Well, sniffed an official of the Food and Drug Administration, which approved this drug in 1995, "It's not an aphrodisiac."

What about men who are perfectly happy with the size, shape, and function of their apparatus? Glad you asked. The technicians (who never seem to rest) have been at work on something you probably do not know you have, namely "sperm stems." It seems that each of us males produce bazillions of teeny-tiny bits of hyperactive matter called "spermatagonical stem cells," the thingamajigs that actually produce our sperm.

So here comes Dr. Ralph Brinster, a University of Pennsylvania veterinarian (Did he say "veterinarian"? Yes, he did), who has been messing with our stems and, through them, found a path to man's long-sought, ultimate goal: immortality. In his lab, the good doctor has discovered that sperm stems can be frozen today and kept frozen until needed, at which time they can be thawed out and (this is not a Monty Python routine, this is true) *transplanted into the testes of a mouse,* where they will promptly begin making human sperm again.

What good is this? Say you are about to die. You want to "live" beyond the grave, so you have a milk carton or so of your sperm stems frozen and stored. Fifty, a hundred, even a thousand years later, a woman could decide (for whatever weird reason) that she wants your baby. No problem—the technicians can unfreeze some of your stems, pop them into a mouse's testes to produce some of your sperm, extract said sperm from said mouse (mouse masturbation being another advance for which American science is so justifiably proud), artificially inseminate the woman, and—voila!—you are reborn.

PEPPERS

Cal Ripken Jr. is a joy to watch as a baseball player. A cinch Hall-of-Famer, still at it for the Baltimore Orioles, in 1995 he broke Ironman Lou Gehrig's supposedly unbreakable record of 2,130 consecutive games played, capping his fabulous night with a home run in his last at-bat. He then took a dramatic victory lap around the field to the thunderous applause of some 48,000 fans in Baltimore's Camden Yards Stadium and to the heartfelt cheers of millions of us watching on television at home.

But Ripken really won my heart by something he did not do. He says he did not watch the videotape of his record, his homer, and his victory lap. "I thought I wanted to look at it right away because the experience was so great," he told sportswriters a year after the accomplishment. "But the longer I didn't watch it, the more I felt the need to just preserve the memory that was in my mind. I have this special perception of it, and I'd like to keep it as long as I can."

The real thing can be so good that it needs no technology to enhance it. Let the natural event suffice. But it seems to be impossible to shoo the engineers off a good thing, even if that thing is the ultimate food of the gods: the noble jalapeño itself, the zenith of pepperdom. For centuries nature's own jalapeño has been a staple of Mexican and Southwestern-U.S. cooking, and now its popularity has swept our continent—I know, because I have been fed *enchiladas verdes* in Milwaukee, *huevos rancheros* in Amherst, *ceviche* in Chicago, and *pescado veracruzano* in Portland—and I am alive to tell about it. All of these comestibles were graced with just the right amount of the pungent, slightly smoky, and moderately hot jalapeño.

But now come the storm troopers of homogenization, the food technologists at Pillsbury, Campbell Soup, PepsiCo, Nestlé, and Kraft, who have used their deep-pocket marketing clout to dominate the grocery shelves and to gain control of more than half of the Mexican-food market in the world, mostly by snapping up regional companies and reengineering authentic products to fit sales strategies. Apparently, to determine the level of "spiciness" in their Mexican edibles, these marketeers are all conducting their taste tests on Norwegian-Americans in someplace like Scandinavianville, Minnesota, where they consider Miracle Whip to be "a little zesty." The new products of these processors are about as Mexican as marshmallows.

Indeed, at the Pillsbury Technology Center in Minneapolis, where engineers actually study things like frijole texture and analyze the porosity of tortilla chips through electron microscopes, a twenty-member development team for its Mexican-food line includes not *one* Mexican or Mexican-American. Christopher J. Policinski, who believe it or not is VP of Pillsbury's Mexican-food subsidiary, essentially says his conglomerate does not care about authenticity, pointing out that "Internally, in our memos, we always put the word Mexican in quotes."

Especially stinging to us Texans, though, is the betrayal of Pace Foods, the once-proud San Antonio company that built a salsa empire on the jalapeño plant. Pace ran a wildly popular commercial a few years back featuring a chuck wagon cook who tried to pawn off a wimpy hot sauce on the cowboys around the campfire. Where was this godawful stuff made? the cowboys demanded. New York City, was the cook's timid reply. *New York City?!* the cowboys bellowed in unison, and then one of them said calmly, but firmly: "Get a rope."

Today, Pace is the property of Campbell Soup, a *New Jersey* company! Get a rope, for the Campbellized Pace has emasculated the mighty jalapeño. Led by Pace's VP of technology, Dr. Lou Rasplicka of Nebraska (need I say more?), the company conducted a top-secret research project in Hawaii to gringo-ize the jalapeño. Code-named "Operation Big Chill" (who comes up with this stuff?), the genetic crossbreeding project literally removed all the capsaicin from the pepper, the natural substance that gives the pepper its kick and its raison d'être in Mexican cooking.

Campbell's food technicians "machined" the jalapeño with a "high-pressure liquid chromatograph" that measures the capsaicin bite. A jalapeño runs about 150 parts of capsaicin per million on average. A real hot one is 400. Campbell's new chili runs zero. *Nada* on the heat index. The result is a pepper that looks like a jalapeño, but tastes like a bell.

Pace's hired apologists say this is about choice, "the freedom of consumers to choose the level of heat in their salsa," the company president told the *New York Times*. Neither he nor the *Times* noted, however, that Campbell's does not need a flock of engineers in white smocks manipulating the genetic heat of jalapeños to make "cool salsa"—just use less of them in the salsa, or none.

Food historian Sidney Mintz of Johns Hopkins University notes that no-heat jalapeños are a primo example of how the technocrats and corporatists of the New World Food Order pretend to embrace a multicultural America, when they actually are deculturalizing our country.

HAPPINESS

C. E. Stubblefield made a name for himself in Texas barbeque, no mean feat in a state where good brisket is worshipped with almost the same fervor as cows are worshipped in India, a state where every Billy Bob and Tammie Jo with a smoker and a jar of "Momma's secret sauce" claim to make the best Q in the known universe, a state where you can find barbequed goat, armadillo, javelina, hormone-free beef, arugula, catfish, ostrich, crawdads, roadkill, tofu, or anything else that might suit your taste.

Stubb, who was taken to BBQ heaven a couple of years ago, served up his specialties at his joint located inside Austin's legendary blues house Antone's, and people came from miles around to enjoy a plate of whatever the man was dishing out that night. His reputation was not in his sauce or smoking technique, not in some marketing gimmick or franchising, but in him. Truth to tell, you could get better barbeque elsewhere, but you could not get a better barbeque experience anywhere else. A tall man filled with tall tales, Stubb and Stubb's (he and his place were one and the same) were pure happiness, just plain fun, and full of good life. When Austin writer Don McLeese asked Stubb what made his food so special, he replied, "I would say it's the people I love to cook for. There's a deep separation between cooking for money and fame, and cooking to make somebody *happy*. My secret is I care about it."

Imagine that. Cooking to produce happiness. Imagine this: organizing work to enhance happiness, governing to promote it, educating to achieve a higher level of it. When asked, nearly all of us say that what we really want from life is "happiness," yet almost nothing in our entire social structure is set up with happiness as a consideration, much less a goal—not jobs, for sure; not schools or careers; not the political system or the consumer economy; not our neighborhoods and shopping centers; not even vacations, which we mostly spend gasping to catch our breath for a couple of weeks before plunging back into the chase . . . for what?

Politicians are practiced at promising a chicken (or a tax cut) in every pot and are adept at asking, "Are you better off than you were four years ago?" (both Reagan and Clinton used that one), but when is the last time you heard anyone in politics ask either you or the country: "Are you happy?"

It is neither a frivolous nor a new question for American politics.

> We hold these truths to be self-evident, that all men are created equal, that they are endowed by their Creator with certain unalienable Rights, that among these are Life, Liberty *and the pursuit of Happiness.*

So reads the second sentence of our country's most basic statement of purpose, the Declaration of Independence, July 4, 1776. The third sentence declares that when a government fails to secure these unalienables for the people, then the people have the right to toss that government and set up another one organized in such a way that is "most likely to effect their Safety and Happiness."

"Pursuit of Happiness" was not merely a throwaway line by Jefferson, but a philosophical concept debated broadly at the time and deliberately chosen by the Founders over "life, liberty and the holding of property," which was favored by some of the more acquisitive Founders. Nor was "happiness" a sly code word for "money"—rather it meant living a meaningful existence, enjoying participation in community life, and having opportunities to develop the whole person, including the mind, the heart, the spirit, and even the funny bone. Sounds right to me, how about you? The Declaration also intended that happiness was for all, not merely for the powerful and privileged. One more thing—the Founders' understanding of happiness also included the community-minded caveat that one's pursuit of it must not encroach upon the happiness of others, a philosophy utterly alien to the modern corporate state.

"Do you know the only thing that gives me pleasure?" asked old man John D. Rockefeller, the profiteer and amalgamizer of the infamous Standard Oil trust. "It's to see my dividends coming in." Now there is a feller with a stunted sense of fun.

It is way past time for our society to reassert America's third *unalienable* right. The political practitioners continue to debate modern-day aspects of "life" and "liberty," but as the national dialogue has become thoroughly corporatized (in politics, in the media, in the schools and elsewhere), "pursuit of Happiness" has conveniently slipped out of public conversation. Any talk of "happiness" today has been perverted to mean consumption, and "pursuit" has been reduced to getting a job, no matter how miserly or unhappy.

There *is* one place, however, where this philosophical issue recently resurfaced in the mass media, albeit too briefly: in the funny pages. Working in the tradition of Walt Kelly's *Pogo,* Bill Watterson's now defunct but still-wonderful comic strip *Calvin & Hobbes* addressed the big issues, including happiness. Calvin is a six-year-old terror, and Hobbes is his "tiger"—a stuffed animal when in the presence of other people, but a "real" tiger and best buddy when he and Calvin are alone. In a 1992 strip the two are cleaning up in the bathroom after a joyous day romping in the outdoors, with Calvin stripping down to get in the tub. "Wow. Look at the grass stains on my skin," the boy says to his pal in the first panel. In the second panel, Calvin then declares to an approving Hobbes, "I say if your knees aren't green by the end of the day, you ought to seriously re-examine your life." Now that is talking my language—unadulterated happiness.

In a four-panel strip in 1995, though, Watterson rendered a corporate perspective on the same matter. Panel one shows Calvin waiting expectantly behind a cardboard box set up as a retail counter, marked: "Happiness 10¢." In panel two Hobbes walks up and merrily inquires, "What do you give people for their ten cents?" Calvin replies enthusiastically, "A water balloon right in the kisser!" Panel three has an incredulous Hobbes asking, "You take their money and then soak them with a water balloon?" "Right," says Calvin. In the final panel, Hobbes asks, "Whose happiness are we talking about?" And an exasperated Calvin, throwing his arms wide and glancing about his cardboard empire, responds, "Who went to all this trouble?"

In our real-life society, the cardboard empire of the corporation has been "enthroned," as Lincoln put it, placing our general and individual happiness at its mercy. Think about it. We allow a single, avaricious special interest to set the conditions of our work, the price of political and governmental participation, the choice of our news and entertainment, the focus of public education, the shape and locations of our "community" enclaves, and much, much more, including the terms of the social contract that exists to bind us into a united society. This is dumber than dum-de-dum-dum. The result shrinks what is possible from our great country, narrows our reach, locks in a widening separation of the few from the many and—most assuredly—produces both a growing unhappiness and, ultimately, an unpleasant uprising by all those who get that water balloon right in the kisser.

Work is a good place to begin the new pursuit-of-happiness debate. Why do we do it? "Work like a beaver," we're told as children, "and you will get ahead in life." But wait a minute—even the most eager beavers only work about five hours a day, mostly at a fairly leisurely pace, and beavers are known to take frequent vacations. Plus, they do work that matters, building sturdy lodges for their family and community, not repetitively soldering a gizmo into a line of gadgets destined for a quick trip to the landfill, or processing a daily load of data that no one needs. Beavers have a life, not a job.

Yet everyone from Bill Bennett (the Reagan Republican who has become the national scold) to Bill Clinton (the Eisenhower Republican who seems to aspire to become Bill Bennett) talks of the "moral imperative" of having a good work ethic and of the "inherent nobility" in having a job. But if jobs represent nobility, why is the corporate state doing all it can to reduce working people to peons, shipping out good jobs, knocking down pay, eliminating benefits, making jobs temporary, and pissing on any remnant of respect for workers that had been built into the nation's social contract over the last sixty years?

In 1996, to show Republicans that he could be every bit as stupidly pious about the work ethic as they are, Clinton proudly tossed America's welfare mothers into the workplace. Fine. Welfare is a bureaucratic nightmare, and most mothers in the system prefer jobs. Except there are no jobs for them, and even if there were, there's no day care for their children and no transportation to get them to work and back. Whoopsie-daisy, said Clinton, who apparently had not counted on complications when he decided to be a moral-imperative Republican. The President of the Greatest Power on Earth then started going around to meetings with everyone from the clergy to corporate chieftains, pleading with them to "do the right thing" and hire these hundreds of thousands of mothers he sent packing from the welfare rolls. They have not. So then, in his 1997 State of the Union address, Clinton proposed *paying corporations* up to $10,000 for every job they offer to a welfare recipient, thereby shifting our welfare money from poverty-stricken mothers to corporations.

Since then, Clinton (along with the Republican Congress and a gaggle of pious governors who likewise have pushed "welfare-to-work" reforms) has boasted till he was hoarse that this tumping over of the welfare safety net has been one of his finest achievements. "Welfare rolls are at their lowest levels in twenty-seven years," Clinton brayed in his 1998 State of the Union address. OK, but did those human beings find work? He and the rest of the pols do not know, nor do they ask. If they did ask, they would confirm what a March 1998 survey by a New York state agency found: Only 29 percent of the 480,000 recipients dumped from that state's welfare programs had been hired into any kind of job, either full-time or part-time. This survey even counted as a "job" any position that paid as much as $100 over three months.

Let us review what our society has achieved by Clinton's exercise in puritanical politics:

1. The President gets to claim that he "put an end to welfare as we know it."
2. Some welfare mothers are getting low-paying, mind-numbing, dead-end jobs.
3. More mothers are getting neither welfare nor work.
4. Corporations are able to get a fat, new subsidy.
5. Other low-income working people must now compete for jobs and wages with this influx of cheap, subsidized labor.
6. We have had demonstrated once again that the work ethic has little to do with work, nothing at all to do with ethics, and everything to do with the corporation maintaining a tight chokehold on working people.

If work does not pay, is not interesting, gets no respect, and is little more than a grind, why work? This is what Panhandle Jack asked himself back in 1927, when he chose to become a hobo. According to superb storyteller and folk singer Utah Phillips, this old fellow hoboed nearly all his long life, having sized up the workplace with this clear-eyed view: "I learned when I was young that the only true life I had was the life of my brain. But if [that's] true . . . what sense does it make to hand that brain to somebody for eight hours a day for their particular use on the presumption that, at the end of the day, they will give it back in an unmutilated condition?"

Granted, not everyone wants to be a hobo. Most of us have other ambitions, especially when we are young. And we have mouths to feed. So making our way and making a living necessarily put us in today's job market. But none of this denies the core validity of Panhandle Jack's observation, as true today as in 1927.

The great majority of us are employees, including most who call ourselves "managers" or "professionals." We take orders. Not 5 percent of Americans have any real say about their jobs. This might seem at first blush a pitiful state of weakness, but doesn't it also suggest a terrific power? We are the 95 percent. The emperor has no clothes! As employees—and more importantly as citizens—we do not have to accept work as it is. Nor do we have to accept that work is all there is.

Marxist thinking? Try Lincoln: "My father taught me to work, but not to love it. I never did like work, and I don't deny it. I'd rather read, tell stories, crack jokes, talk, laugh—anything but work."

We are not taught it in school, but Old Abe was hardly a loner in his slim regard for the work ethic, which is now handed down to us with thunderous absolutism, like God handing down the tablets to Moses. Remember, though, even God rested, and if you think that this Creator had no sense of fun, I can only ask you, Why do men have nipples?

Even those hardy immigrants from the Old Sod who came across at great risk and found their way to the land of bootstrapism through Ellis Island were hardly the unblinking worshippers of thriftiness, punctuality, hard work, obedience, and determined upward mobility presently portrayed by the powers that so eagerly want this myth to be our sole role model. In the 1880s and 1890s, at the height of the rise of the robber barons, the Victorian work ethic was nowhere near as popular as immigrant "saloon culture," which was not simply about drinking, but about family, friends, enjoyment, community . . . happiness. As the fine and fun historian Richard Shenkman records: "It wasn't that the immigrants were lazy. It's just that they thought there were often better things to do with one's time than work."

Despite the efforts to bury this spirit, notice that the most common phrase in today's workplace, no matter how tall the building, is TGIF! And for all the corporate culture's insistence that everyone get ready for the exciting twenty-first-century global workplace, do you ever hear people shouting TGIM?

It is time for the thirty-hour workweek in America. Break it into five days of six hours each, or four seven-and-a-half-hour days, or whatever combination works for you. It will work for the economy, too (as the shorter workweek has done for Germany), increasing everyone's productivity, shrinking both unemployment and underemployment, and putting more consumer dollars to work at the grassroots of our economy (after all, money is like manure—it only works if you spread it around). It is also another way, besides wage and salary increases, for working folks to begin sharing in those economic gains they have been producing, which now flow almost exclusively to the most privileged investors and to the CEOs.

Even more important, less-work/more-workers allows us to begin taking back our lives. Work is killing us, physically and spiritually, because we spend way too much time there. It is killing the family and the community, too. The forty-hour week? Gone. Unionists of yesteryear fought and died for this social landmark (the bloody Haymarket Riot of 1886 was about shorter hours, for example), and the forty-hour limit finally was achieved in the 1930s. But in the past couple of decades companies have relentlessly been inching up their time demands, and families, having to scramble for more hours to compensate for falling wages, have had no choice but to go along. The forty-hour week has become a joke—it now averages fifty for wage earners, and it is common for folks to be in the salt mines sixty hours a week.

By the way, whatever happened to the lunch hour? Only 12 percent of us get a full hour or more for lunch these days. When companies downsize, they do not reduce the amount of work to be done—just the number of employees to do it. With this heavier workload, something has to give . . . and one thing to give is the time for a midday interlude. Our average lunch break is down to *twenty-nine minutes*. Four out of ten Americans say they really don't take a lunch break at all—they work right through it or grab a quick snack at the desk.

Economist Juliet Schor, author of *The Overworked American,* finds that the typical employee is on the job about twenty more workdays in the 1990s than in 1969, not counting the commute to and from work, which is up significantly, adding not only to the hours, but also to that always-important Pissed-Off quotient. This is the same period during which eight out of ten Americans have seen their pay levels drop. Less income for more hours. Economists have a technical term for this: "stupid."

The increasing time strain is especially tough on parents who have young kids. With wages tumbling, both Mom and Dad are compelled to be in the workplace nowadays, just trying to make ends meet, and both of them have noses-to-the-grindstone more hours than the average worker does. Their younger children are in day school, the older ones are latchkey kids. Mom and Dad are half-dead by the time they get home, and dinner is a quick run through McDonald's or some such before packing the kids off to bed and getting ready for another day.

What a joke it was to see Bill Clinton and George Bush standing side by side at the White House in 1997 to announce that they were launching a special, bipartisan campaign to "promote more volunteerism" among the American people—two weakling Presidents who have had their heads stuck up their asses for the previous eight years while their corporate funders ran rampant over the middle class, leaving a majority of people so wrung out with work, low wages, and worry that they do not have a spare minute or an ounce of energy left to volunteer for anything.

High tech has only increased family stress, because now the boss can have employees working nights and weekends on their home modems. Last winter, for example, my town of Austin had a rare two-day winter storm with a snow-sleet-ice combination that flat shut the place down—no school, no shops, no nothing. Except work. "Personal computers . . . enabled many Central Texans to work from home Monday without worrying about icy conditions," marveled a business writer for the local paper.

Now comes the global economy! Work, work, work. Train, train, train. Get in line to cross that bridge to the twenty-first century, says our President, or you'll be "a loser" (his actual term), adding that we must "adapt to the exciting new world market" (his actual phrase), in which we must best our competitors, even if it kills you. But look!

There is one of our competitors—Germany. German factory workers are paid an average of $6.50 an hour more than American workers get, and Germans work 348 hours a year less than we do. Why don't we compete with that? Also, German employees average five weeks of vacation time every year. In Austria, they get six weeks. The Irish get three weeks. We average two.

Americans *want* more time off. An official with the Roper polling outfit told the *New York Times* that "downsizing of industry did something fundamental to people's thinking about where they get satisfaction in their lives. There is a decline in the belief in work as a source of satisfaction." There is a surprise, eh? A majority of us now tell pollsters we would even sacrifice some wage increases in exchange for more time to be with our families and pals, to pursue other interests, or, hell, as Panhandle Jack advised, just keep our brains to ourselves for a while.

The thirty-hour week goes to the core not only of *how* our society organizes the workplace, but also of *who* organizes it, and it asks a bigger question, too: Who owns our time? The robber barons said they were the sole organizers and owners of time, but after years of struggle, working folks won a New Deal sanctioning in law our right to have a say. Now, though, the robber barons have crept back.

Michael Ventura, whose column in the *L.A. Weekly* is always a thought-provoking read, wrote an especially powerful piece a few years ago about his life in the job game and how he finally figured out that by getting two weeks off each year, "it would take me 26 years on the job to accumulate one year for myself. And I could only have that year in 26 pieces, so it wouldn't even feel like a year. In other words, no time was truly mine. My boss merely allowed me the illusion of freedom, a little space to catch my breath in between the 50 weeks that I lived but *he* owned." Ventura tells of his own Panhandle Jack epiphany: "Slowly, very slowly, I came to a conclusion that for me was fundamental: My employers are stealing my life."

2

CLASS WAR

LET'S CHECK THE *DOUG* JONES AVERAGE

- ★ HOW YA DOIN'?
- ★ STATISTICAL WHITEWASH
- ★ THE LION'S SHARE
- ★ HOW DO YOU SPELL "BOSS?"
- ★ RUTH'S BRICKS
- ★ ST. PETER
- ★ A SHINING CITY ON THE HILL
- ★ THE NEXT BEST THING TO SLAVES
- ★ LABOR DAY
- ★ EARRINGS ON A HOG

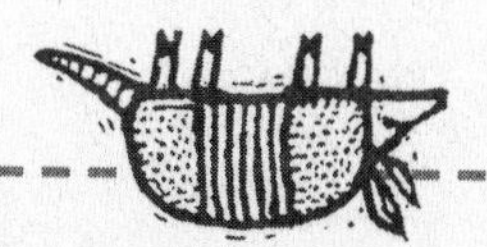

HOW YA DOIN'?

John Kenneth Galbraith won my deepest respect years ago when I heard him say: "If all economists were laid end-to-end—it would be a good thing."

Economists are the voodoo priests of America's established order. They are kept well-fed by Washington and Wall Street in exchange for being brought out of their dark warrens periodically, beady eyes blinking in the glare of daylight, to assure us that "America is prospering," even though we Americans are not. Think about that concept. We are being asked to believe the economy is something separate from us, something more important than the people in it, something we serve, rather than something that serves us. If that doesn't make sense to you it's because it *doesn't* make sense.

The amazing thing to me is that people who laugh at Gypsy fortune-tellers take economists seriously. Our public officials seem befuddled by the voodoo, basing our country's economic policies not on what would be good for the people, but on what they are told would be good for the "economy." In practically every case, what's good for the economy turns out to be (big surprise!) what's good for Wall Street. The media buy into the voodoo, too, monitoring the Street's Dow Jones Average as though it is America's heartbeat. Just you try to get away from the "Dow"—it's on every TV newscast, it runs on radio more frequently than ads for bug exterminators, it's in every daily newspaper and makes the front page more often than floods, blizzards, and hurricanes. We are constantly bombarded with the Dow, when what we need to be concerned about as a country is the *Doug* Jones Average—how's Doug doing, what's the price of Spam this week, how's the short-term inventory of durables down at Big Bubba's Pay-and-Go used-car lot?

What is this Dow Jones Average anyway? "Dow Soars Higher Than 9,000" blare the good-news trumpets of the media, telling us that the economy is "whizzing!" But higher than 9,000 what? Some of my ex-hippie neighbors in South Austin have been higher than 9,000 since 1968, but does their altitude tell us anything we need to know?

In reality, the Dow is an insignificant indicator of what is happening on Wall Street, much less on your street. It is an index of the performance of only thirty of the 2,600 stocks listed on the New York Stock Exchange, and all the thirty are such blue-chippers as AT&T, GM, and IBM. Yet politicians in Washington monitor the Dow religiously, as if Moses himself were handing down hourly updates. When Bill Clinton jettisoned his campaign pledge to launch a much-needed jobs program for people, he said he did so because he did not want to upset the market, and the next day the Dow applauded his broken promise with a nice little uptick in its average.

The Dow is like that—perverse. When you do poorly, it shoves its statistical fist in the air with an exultant "Yes!" One sure predictor of a happy Dow day is the announcement of a huge corporate firing: "AT&T to eliminate 40,000 jobs," the *New York Times* reported on January 2, 1996. "Dow Jumps 60 points," read the next day's headline.

These days, whenever a company dumps another load of workers onto our streets, it can expect Wall Street speculators to show their approval by shouting "way to go," immediately bidding up that particular company's stock price. For example, just in the first three months of 1998, AT&T's stock roared to its highest level in a year following news that 15,000 more of its employees were to be fired; JCPenney shares shot up $1.31 the day it revealed plans to dismiss 4,900 employees; Black & Decker stock climbed $4.50 on news that 3,000 of its jobs would be eliminated; Chase Manhattan stockholders gained $6.43 a share when the bank announced 2,250 terminations; and the Money Store, Imation Corporation, Lam Research Corporation, and United Technologies also were among the companies firing employees and getting a giddy Wall Street goose in their stock prices.

This phenomenon goes from being coincidence to being dastardly when you realize that the top executives of major corporations now have most of their own compensation tied to their company's stock value. What we have here is a powerful incentive for executive suite mischief—the more workers those executives fire, the more the company's Wall Street price begins to whiz, and the more the executives get paid.

Sure Wall Street is whizzing, it's whizzing on you and me.

This is why I propose that, instead of the national fixation on Wall Street's Dow Jones Average, we tune in to the *Doug* Jones Average, a real-life indicator of "How ya doin'?" for the 60 percent of Americans who don't own any stocks and bonds. (Faint clatter of teletype machines heard in background, as the quick-paced, modulated monotone of the Doug Jones reporter begins):

> The Doug Jones Average took a tumble today in a flurry of heavy trading in the *Job-Futures Market.* Major industrials, including AT&T, Burlington Northern, and Apple Computer announced more than 97,000 additional employee terminations in January, causing market analysts to predict that last year's record level of 3,100 job losses each workday would be exceeded this year. There was a brief uptick in the market at midday over news that a ball-bearing company in Euchrys, Ohio, was adding new production workers, but the rally sagged on later reports that only twenty jobs were available, and the futures market closed with losses badly outdistancing gains. Especially sobering to the Doug was a late report that Eastman-Kodak has dismissed 10,000 more employees, including its chief executive, terminated because Kodak's board of directors felt he was not firing employees fast enough.
>
> Checking today's *White-Collar Anxiety Indicator,* we find it still in the Red Zone at 117.9 and trending upward as insider information from IBM has revealed that Big Blue is forcing its software employees to train recruits from India to do their jobs, with expectation that these high-tech positions will go abroad to the facility IBM has built in Bangalore, India.
>
> Turning now to our daily *Lunch-Bucket Index,* profit-taking on BLTs is causing a consumer retreat. While tomato prices declined due to seasonal supply and a surge of Mexican imports under NAFTA, lettuce prices gained markedly, bacon is up nearly five points, and white bread is up, too, by three-point-seven. Beer was relatively flat at a point-two increase. On the homefront, the Doug's bottom line was pinched by a three-point increase in rent, a four-point-eight jump in garbage collection, and a seven-point zap in cable TV rates.

This just in from our *Hollywood Hotline:* Disney Inc. has announced that its latest sequel to *101 Dalmatians* will be downsized. In an austerity move, fifty of the spotted canines are to be eliminated from the movie script altogether, and the other fifty-one are to be replaced by Mexican Hairless Chihuahuas. The tiny Mexican dogs will have to have their appearances enhanced by makeup and computer graphics, but Disney CEO Michael Eisner assured skittish investors that the added technological costs will be more than offset by the fact that Mexican Chihuahuas work for pesos and require only 10 percent as much dog food as Dalmatians.

To close on a bright spot, the *Bonds Market* continues to pay major dividends, as advances outpace declines. Barry's batting average fell eight points in the past month, but is still a robust .344, while his home run output is up by ten and his 82 RBI now leads the major leagues.

That's today's Doug Jones Average, for those of you worried more about the price of socks . . . than stocks. Remember, it's not the Dow Jones Average that matters—it's yours.

STATISTICAL WHITEWASH

All you really need to know about official statistics is that the Hundred Years War lasted 116 years.

But, hey, who's counting? (OK, the people who died during those extra sixteen years might want to be counted, but picky-picky. Statistically speaking, they were within the margin of error.) Besides, that war was way back in the fourteenth and fifteenth centuries, the Dark Ages for number-crunching. Today, using the sophisticated techniques of statistical science, using those big-honker F–150 LX Crew Cab computers to process the data, using the Hubble Telescope to peer light-years into space and see the war *before* it happened, using actual Ph.D.'s with spinning beanie-cap propellers to make the chronological calibrations—our proud U.S. Bureau of Great Big Numbers (acronym: US Bureau of Great Big Numbers) is able to say with absolute precision that the correct reference should be: the "Hundred Years (Approximately) War."

They are even less accurate in counting the victims of today's class war. Although every morning's paper brings us news that some other outfit with a name like MegaMultinational Inc. has drop-kicked another 10,000 or so employees through the goal posts of global greed, the nation's official job statistics stay magically cheerful. This is especially useful to Wall Street and its toadies in Washington, because it gives them the political cover they need to fend off any serious effort to do anything to stop the MegaMultinational Incs. of our country from mugging the middle class in broad daylight. They use employment statistics like the drunk uses a lamppost—not for light, but for support.

Each month the Bureau of Labor Statistics rolls the drums, rolls the dice, and rolls out two official government numbers that are used to cloak the muggings. First is the ever-popular, widely ballyhooed stat that you often see as headline news: "Unemployment Rate Stands at Only 5.4 Percent."

Only? There are more than a million, three hundred thousand living, breathing, hard-hit human beings behind each percentage of unemployment, so we're talking about nearly seven-and-a-half million of our neighbors and family members caught up in that diminutive datum.

They are only the tip of the iceberg. Millions more of the unemployed don't even become statistics. The BLS comes up with its official monthly count by simply calling 12,000 households and, in effect, shouting: "Hey, is everyone in there either 'employed' or 'looking for work'?"

Not much more nuance is allowed. For example, let's say you've been down and out and trying to find a job for a long, long time—so long that you've given up the hunt, at least temporarily. "Are you employed?" the caller asks. No. "Are you looking for work?" Not this month. So in the strict statisticalese of BLS, you are not included in the monthly jobless picture, even though you are jobless. The agency does have a name for you though: "Discouraged Worker." Isn't that genteel? There are about half-a-million discouraged folks out there each month, but they don't count with the unemployment counters.

Let's keep a running tab: 7.5 million or so "officially" unemployed plus a half-million or so "discouraged" workers equals 8 million actually unemployed, so far.

Let's also add in those of you who are down with the flu when the BLSers call. You are in the market for a job, but there you are in bed when they call so you are not technically out "looking for work," are you? You don't get counted, either. Or maybe you do not have a job and are looking for one, but in a burst of candor, you say to the surveyor, "Well, I did do a little work on my brother-in-law's house last week and got a few bucks for that, but . . ." But nothing, bucko—as far as the BLS is concerned, that puts you among the gainfully employed and drops you from the bad-news bin into the good-news group. The government definition of "unemployed" is so narrow that, until some bad press forced BLS to change its policy recently, those who said they had been "laid off" were counted as still being employed. The bureaucratic rationale was that laid off is just a kind of "suspended employment" and your company would soon be laying you "on" again—totally ignoring that, in the vernacular of today's harsh workplace, laid off means you are kaput, gone, out of here, fired, toast . . . unemployed.

No one knows for sure how many of the unemployed slip through these definitional cracks, but a conservative estimate figures that more than half-a-million a month do. So this raises our tally to more than 8.5 million Americans with no jobs.

Then there are you proud ones—especially you middle managers, accountants, engineers, and other college-educated, highly skilled white collars who suddenly find yourselves holding a pink slip. For the first time in your twenty-five- to thirty-year career, you have no office to go to, and no other company is returning your calls. Are you unemployed? "No," you tell the BLS surveyor, "I'm *self*-employed." Sure you are, and there are now more than 3 million of you in your spare-bedroom "virtual offices" trying to look busy and praying that someone will call and offer you a real job. Add this 3 million to the 8.5 million above, and we are nearing 12 million in the dismal ranks of the unemployed.

Still, this is not the half of it. Have you chatted up a waitress recently? Chances are she has at least one college degree and is one of some 20 million Americans who have settled for lower-paying jobs that make no use of the education, expertise, and experience they worked so hard to gain. In this category also are employees hired as "temporaries" to do the same work as "regular" hires. The difference is that the temporaries get lower pay, receive lesser or no benefits, and generally are treated within the corporate caste system as scuz. Some companies actually require these untouchables to wear different-colored employee badges that announce their inferiority. Microsoft is one such company, with as much as a third of its Seattle-area workforce being temporaries—and being treated like illegitimate children at the family reunion. A few Microsoft examples, as reported by the *Seattle Times*: temporaries are not invited to the annual picnic, nor can they attend such company morale boosters as after-hours pizza parties and dances; temporaries do not get such employee perks as health-club memberships and the opportunity to purchase company-made software at discounts from the company store; and temporaries are prohibited from joining Microsoft's athletic teams or even from playing on the company's athletic fields when the regulars are not using them. Although these jobs are designated "temporary," implying a short-term employment condition with an opportunity to move up, it's not uncommon for Microsoft and other companies to hold employees in their temporary status permanently. There even is a new term for them: "Permatemps."

In addition, about 4.5 million more of us work part-time but want and need full-time jobs. Also, toss in the 25 million people (one out of five workers) with "jobs" that pay under six bucks an hour. Nearly half of these are minimum-wage employees, which means trying to make ends meet on barely $10,000 a year. This is not a job, it's wage slavery. Even adjusting these categories for overlaps, it still means there are around 25 million chronically *under*employed and underpaid people out there—people whose job needs are not being met by our economy. Add them to the 12 million jobless, and we've uncovered about 37 million people—plus their families—who grind their teeth every night over the lack of decent jobs.

Bottom Line: The relevant statistic for assessing (and addressing) our country's job health is not Washington's trifling 5.4 percent, but the much grimmer reality that a good *third* of America's workforce is mired in unemployment and permanent underemployment.

Yet our politicians, pipe-smoking editorial writers, and even Presidents (no, make that *especially* Presidents) blithely embrace the official stat and continue skipping merrily down life's primrose lane, trying to put a Happy Face on an economy that most Americans know is not working for them.

The blissful complacency of politicos and pundits today is reminiscent of Calvin Coolidge's in the 1920s, back at the dawn of the Great Depression. Asked about rising unemployment across the countryside, Silent Cal shrugged and said: "When more and more people are thrown out of work, unemployment results." (I realize that Coolidge was not the brightest bulb ever placed in the Oval Office, but surely someone was fooling with his dimmer switch when he said that.)

The official unemployment stat is only one rose-colored lens worn by Washington, Wall Street, and the media when looking at the economy. The second lens is yet another monthly product of the BLS: the number of new jobs created. This one is particularly dazzling to occupants of the White House, who are always eager to report that their economic policies are doing a world of good. It was this job-creation statistic, for example, that did so much to make George Bush what he is today: a one-term ex-president. You might remember that Bush kept insisting in 1992 that "that economy thing" is "great guns," "Uptick City," and "don't cry for me,

Argentina," which was Bush-ese for "Why is everyone so darn glum about the doggone economy when I've created hundreds of thousands of new jobs for them?" In fairness to Bush-the-Ex, this is exactly what the BLS job-creation statistics were showing, leading his advisers to assure him that the economy was humming. Unfortunately for poor George, the statistics bore no relation to the real-life experiences of the voters, and the more he cited the BLS, the more he sounded like he was full of BS.

Oddly, though, the man who capitalized on Bush's gullibility has stepped into the exact same statistical cow patty. Within a year of becoming President, Bill Clinton was crowing about the number of new jobs he had created, and he has continued to crow about it ever since. In his 1997 State of the Union address, for example, Clinton climbed up on the peak of the national barn and let go with this cock-a-doodle-doo: "Over the last four years, we brought new economic growth . . . creating over 11 million new jobs."

Take a cold shower, Bill. Maybe you believe that stuff, but we know better. As Will Rogers once put it, "It isn't what we don't know that gives us trouble; it's what we know that ain't so." Common sense should tell Clinton that when corporations are in the news every day shedding huge numbers of jobs like a mangy dog sheds clumps of hair, maybe he should begin questioning the accuracy of the job figures he's fed each morning.

Time for a reality check. The BLS calculates how many new jobs have been added in a month by taking another survey, this one a random sampling of about 10 percent of America's companies. Aside from the legitimate question of whether this small snapshot represents anything approaching an accurate picture of what really is happening job-wise, BLS totally skews its own monthly number by engaging in a jaunty bit of hocus-pocus it calls the "bias factor." What it does is assume that its survey mostly catches larger companies, thereby overlooking jobs that small outfits might be creating. To correct for this presumed bias, the agency simply—*abracadabra!*—adds a hundred thousand jobs or so to its official count each month, jobs that it *guesses* are being created by small businesses, even though it cannot find these jobs in the real world. They cannot be found because they are not there. Small business organizations like the National Federation of Independent Businesses report that far from adding jobs, most of their members have been letting employees go.

Another bit of queerness in the BLS stat is summed up by the story of the guy in a cafe, sitting on a stool at the counter, reading the morning paper and drinking coffee: "Here's a report that Bill Clinton has created thousands of new jobs," he says to the waitress. "I know," she responds dryly, "I have three of them." The survey makes no distinction between a real job and the part-time, temporary, low-paying "McJobs" our economy foists on us today. If you are one of the millions patching together two or three part-time gigs to make ends meet (maybe you run a 5 A.M. paper route, work 9 to noon as a paralegal for some law firm, then grab the afternoon shift at a 7-Eleven)—congratulations! BLS counts you three times.

On April Fools' Day, 1994, for example, President Clinton hailed the remarkable news from the BLS that 456,000 new jobs had been created the previous month. But John Crudele, the astute financial columnist for the *New York Post,* dissected this salutary statistic and found that 112,000 of the jobs—nearly a fourth of the total—did not exist; they were phantoms produced by the bureaucracy's "bias factor." More than 300,000 of the rest were part-time positions.

Take a more recent example: "Payrolls Swell by 271,000," trumpets the *New York Times* in a February 8, 1997, headline, citing the BLS job survey for January. "Job Figures Point to Vigor in the Economy," announces the front-page headline in the *Los Angeles Times* on the same day. These and other stories use words like "robust," "rousing," and "enviable" to break the joyous job news, with one corporate economist squealing, "Pinch me because this just seems too good to be true."

It was. Aside from the usual phantom jobs bloating the bureau's numbers (a bloat not reported by any of the media), practically all the actual new jobs counted in the January survey were low-paying, no-benefit service jobs, with an especially big surge in temporary work. Indeed, the BLS report noted cryptically that the big job jump was magnified by "weather-related factors." Such as? Snow. It seems that an Arctic storm swept the country in January, even plunging deep into the South. The result was a major increase in what your Ph.D. economists technically refer to as "jobs shoveling snow."

Presumably this activity was the source of the economic "vigor" hailed by the media.

What we have in America's job numbers is an example of the statistical art of drawing a mathematically precise line from an unwarranted assumption to a foregone conclusion. It's the biggest whitewash since Tom Sawyer's fence.

THE LION'S SHARE

Being fired is not funny and firing someone is not fun. I've been on both ends of the gun.

But it has been at least slightly amusing these last few years to watch corporate PR flacks devise ever more convoluted phrases to try cloaking the cold fact that their companies dump millions of good employees on the scrap heap of economic "progress." In the media, "downsizing" is the preferred euphemism for this phenomenon, but it is nowhere near as inventive as these other dollops of doublespeak applied to the tidal wave of mass firings:

- "Release of Resources" (BankAmerica).
- "Competitive builddown."
- "Career-transition program" (GM).
- "Employee outplacing."
- "Schedule adjustments" (Stouffer Foods).
- "Reengineering."
- "Management Initiated Attrition" (MIA for short).
- "Normal payroll adjustment" (Wal-Mart).
- "Negative hiring."
- And my favorite, "Decruiting."

By whatever name, the motivation behind this assault on the middle class is nothing more noble than old-fashioned greed. Company executives kick out and knock down so many because Wall Street investors and their own boards of directors reward them lovingly for doing so, and because they now realize no outside force will raise a hand to stop them—both political parties are in their pockets, government regulators are sheep in wolves' clothing, and unions have been roped, throwed, and hog-tied. Near my home, there is a drive-through liquor store with the impatient name of Git It & Go, and that fairly well expresses today's corporate attitude, grabbing all they can as fast as they can, abandoning any ethical sense that their destiny is or should be linked to that of their employees or communities.

What they seem intent on taking is the "lion's share," and, as we're told in Aesop's fable, the lion takes not merely the largest part of it, but all of it:

> The cow, the goat and the sheep went once a-hunting with the lion, and they caught a stag. And when they came to have their part and share in it, the lion said to them:
>
> "My lords, I want you to know that the first part is mine, because I am your lord; the second because I am stronger than you are; the third because I ran more swiftly than you did; and whosoever toucheth the fourth part, he shall be my mortal enemy."
>
> And thus the lion took for himself alone the stag. And therefore this fable teacheth to all folk that the poor ought not to hold fellowship with the mighty. For the mighty man is never faithful to the poor.

HOW DO YOU SPELL "BOSS?"

"Little bee sucks the blossom," goes the old Western-swing song by Bob Wills and the Texas Playboys, "but the big bee gets the honey/Little man picks the cotton, but the big man gets the money."

This reality was never starker than it is today—not just in the cotton fields, but in the corporate fiefdoms of the Fortune 500.

In Austin, Texas, where I live, there was a legendary bar in the late 1970s and 1980s called Another Raw Deal, run by two characters named Lopez and Fletcher. Serving longnecks and bodacious french fries, the Deal was a hangout favored by writers, musicians, lawyers-who-used-to-be-hippies, maverick politicos like myself, and assorted other riffraff. (With such a clientele, the bar's washrooms were graced with top-of-the-line graffiti, including this political offering in red Marks-A-Lot next to the chrome blow-dryer: "Push here for a message from Senator Phil Gramm.") This roadhouse was a bit of Austin heaven—a laid-back, beer-soaked, grease-smoke-in-the-air kind of place where eccentrics were normal, iconoclasts were in the majority, music and talk were boisterous, drunk was OK, and excess was applauded. The Deal's slogan was "Too much is not enough."

Imagine our shock and dismay, then, to learn a decade later that our slogan—an unfettered whoop of defiance against the established order—had been expropriated and perverted into a rigid canon of greed by the established order itself, by guys in Gucci shoes and Pucci ties who specialize in mugging working-class people.

By the nineties, Austin's Raw Deal was gone, but America's raw deal had only gotten started.

Do you make $282,000 a year? Most of us do not. Less than 1 percent of Americans make that much, but one who does is Michael Eisner. He's head Mouseketeer at Disney Inc., and his pay in 1997 was $282,000—not for the year, not for a month's work, or a week's or even a day's. He was paid $282,000 *an hour.* Plus a car. This does not even make Eisner the fattest dog on the porch. Consider Bill Gates, geek-in-charge at the monopolistic empire of Microsoft. According to Internet journalist Brad Templeton, who maintains a Web site called "Bill Gates Wealth Index" (http://www.templetons.com/brad/billg.html), Mr. Microsoft has been raking in money at the rate of $500,000 an hour, or about $140 a second. Templeton says we should think of it this way: If Gates was striding down the sidewalk and dropped a $500 bill, it would not be worth his time to stop, bend over, and pick it up. That four seconds would actually cost him money! Eisner and Gates exemplify the wretched excess in today's executive suites, where pay packages begin at a million a year, and it can take more than half a billion dollars to be king of the hill, to rank as the year's highest-paid chief executive.

Not so very long ago, a CEO was considered merely important, much like the lead mule is important, keeping all of us other mules on the straight and narrow and pulling the wagon. In 1960, at the height of America's industrial might, the chief executive averaged $190,000 in pay—about forty times what the worker on the factory floor made.

Today, though, CEOs have refashioned their image into economic purebred stallions who must be pampered, stroked, and fed only the finest, sweetest, richest alfalfa—separating themselves entirely from the mule team. They have reconfigured the corporate structure to establish themselves as our new moneyed elite with wealth, power, and perks that royalty would envy. CEOs now average $7.8 million in pay—*326 times* what workers down on the factory floor are paid.

And there's the rub. It is this disparity that matters, not the fact that those at the top are getting rich (as Mark Twain put it, "I am opposed to millionaires, but it would be dangerous to offer me the position"). The chiefs no longer want their share, but the "lion's share," and they are willingly shoving everyone else away with one hand while they use the other to grab more and more for themselves.

Meet Robert Allen. In the early nineties, while he was top boss at AT&T, he engineered a takeover of NCR computer company, which turned out to be a beaut of a business blunder. It cost the phone giant a few billion bucks—and cost 8,500 former NCR employees their livelihood. Did Allen get booted, docked in pay, scolded, or what? Bingo if you chose "what." He got a 41 percent pay hike in 1995, lifting him to nearly $5.5 million a year on Wall Street's greed-o-rama chart. In 1996 Allen took the bloody ax to his own company, lopping off the heads of 40,000 AT&T employees in one blow, a stroke that swelled his own wallet with another multimillion-dollar pay hike.

The *Wall Street Journal*, ever solicitous about the care and feeding of CEOs, moaned in April 1997 that poor little Bobbie Allen had seen "his 1996 bonus slashed 18 percent," due to the company's "disappointing performance." The *Journal's* reference was not to the disappointment of the 40,000 axed employees, but to the stockholders, who saw their gamble on AT&T take a 9 percent dip in '96. Before you reach for a tissue to dry your eyes over Allen's "punishment," do the rest of the math: His base salary for '96 actually ticked up a bit, to roughly $1.2 million; add his bonus of $1.25 million (down 18 percent, as reported); toss in $2.2 million he drew from his "long-term incentive plan," $1.3 million from his existing stock options, $875,475 in "other compensation," and about $1.3 million in new stock options he was awarded . . . and you could write quite a different headline about Allen's total compensation in 1996, which came to $8 million plus change—a pay *raise* of 45 percent. When *Newsweek's* Allan Sloan confronted him with the ethical pigginess of enjoying a $2.5-million personal pay increase during the year he whacked so many other human beings in AT&T's employ, Mr. Allen responded with a rhetorical question: "Is it fair? Hell, I don't know. I don't make the rules."

But *of course* he makes the rules, and just you try to change even one of them. Every time some shareholder, consumer group, union, member of Congress, or whoever has the temerity to propose the slightest trimming of the totalitarian power that CEOs wield in our economy (including Allen's power to profit personally by inflicting gross pain on 40,000 of his firm's employees), AT&T unleashes the full force of its lobbyists and lawyers, PACs and PR flacks to crush the rebellion and preserve the status quo. These guys have the ethics of grave robbers.

Meet Lou Gerstner, another CEO who has built a following on Wall Street because of his fondness for firing IBM employees. In the name of streamlining Big Blue, as the once-respected computer company was known to its loyal employees, he booted some 200,000 of those employees out the door in '93 and '94. But somehow his streamlining edicts never reached the executive suite—Gerstner tripled his own pay in '93, and by '94 he was wallowing in an annual salary of $2 million, a nice bonus of $2.6 million, and $7,755,000 more in other compensation. This $12-million-a-year "streamliner" is a lot like the submarine commander in a comedy bit that Bob Newhart used to do: The commander was addressing his crew just before they embarked on an especially dangerous mission, telling them in stark detail about the firepower and ferociousness of the enemy they would soon encounter, conceding that their chance of survival was nil and that their deaths could be gruesome. Still, he urged them on for God and country, concluding by saying, "Men, my only regret is that I personally will not be able to come along with you on this mission."

The troops must have stared at that commander the same way IBM employees looked at Gerstner when they learned not only about his pay package, but also that he was retaining all nine of the company's private jets to scoot him to various destinations around the country, including trips to the three golf courses and country clubs IBM maintains for him and other executives. Especially galling to employees who had been fired just before Christmas was the revelation that Big Blue spent a ton of cash for a float in the Rose Bowl Parade. As one employee who had been cashiered after twenty-eight years expressed it: "They say they don't have any money and then they drop a quarter of a million dollars to buy a big flower pot."

The icing on Gerstner's cake came in '95, when he did decide to make some cuts at the executive level—not in the pay of executives, mind you, but in the pay of their secretaries. Whacked them back by a third. The same week, he turned around and hired a $118,000-a-year gourmet chef for his executive dining room. After all, a man works up a hell of an appetite swinging a budget ax.

Meet John W. Teets. He heads Dial Corporation, maker of such products as Dial soap, Breck shampoo, and Brillo cleaning pads. This gentleman prides himself on his corporate cost-cutting prowess, unflinchingly slicing hundreds of employees at a time from the payroll and decreeing that everyone who remains must lead a Spartan existence. Except for him. Teets built a red-granite, twenty-four-story Taj Mahal of a headquarters building in Phoenix, far removed from Dial's factories and rank-and-file employees, and far above the penny-pinching pieties he preaches. This fellow is a living example of the old adage: The higher the building, the lower the morals. His own top-floor office boasts marble floors, mahogany paneling, antique furnishings, and Chinese rugs. Asked by the *New York Times* about this posh environment, Teets feigned modesty and pointed at the digs of another corporate big spender: "I like Ronald Perelman's office better—he's got more antiques."

Antique envy aside, Teets can sit high atop Phoenix in his finely feathered nest, preening himself and counting the eight-and-a-half-million dollars a year he takes in pay. And when he grows weary of the Sun Belt, he can summon one of the jets the company maintains for him and fly off to the bright lights of the Big Apple, where Dial keeps a well-appointed apartment for him overlooking Central Park.

Teets has been especially stern about the way employees use company funds—he wouldn't want a dime diverted for a personal call on the office phone, for example, or any paper clips being taken home by a secretary. Yet, in 1995, he got Dial to put $10 million of the corporation's money into the Arizona Diamondbacks, Phoenix's new major-league baseball franchise in which Teets happens to have a personal stake. Asked about the ethics of funneling ten million Dial dollars into his personal venture while he cut jobs and demanded austerity from others, Teets regally responded: "I am retiring in a couple of years and I've got to be thinking about what I'm going to be doing. We're talking about futures here."

Messrs. Allen, Gerstner, and Teets are typical of the aloofness, even imperiousness, that characterizes most of the elites who now rule America's corporate economy. Michael Eisner, for example, the $100,000-an-hour man, chopped the health-care plans of the low-paid minions who work in Disney's theme parks, apparently oblivious that his own extravagant pay was more than the annual health-care premiums of all those employees combined. Nor did Eisner seem fazed by the stunning hypocrisy of gutting their meager medical benefits while protecting every bit of the gold-plated Cadillac coverage the company provides for him and his loved ones.

This is one reason that so many at Disney and elsewhere have noted that "boss" spelled backwards is double-S-O-B.

It is common to assail members of Congress and other Washington politicians for being out of touch with everyday reality, but they are regular Joe Six-Packs compared to these CEOs. Check out Alexander Trotman, big boss at Ford, who has never in his entire career as an auto executive bought a car. He has never experienced such joys as hassling with a sales rep on price, trying to decipher the warranty, getting the thing insured and inspected, making payments, taking it in for service and getting one of those blow-your-socks-off repair bills. At Ford, you see, a car just "magically" appears for those on the executive track—it is always gassed up, tuned up, washed and waxed, and has that "new-car smell." And when the newness begins to wear off, another new one "magically" appears. Trotman told the *New York Times* that while he has never personally been a customer, "I know a lot of dealers, and I don't see why I have to buy a car to get in touch with the dealer-customer relationship."

Even in the rare cases when CEOs get bounced, the experience is one that an average working stiff would beg to have. The platinum parachute of N. J. Nicholas Jr. illustrates the point. Ousted in 1992 as president and co–chief exec of Time Warner because of poor performance, he received:

- A lump-sum severance payment of $15.8 million.
- A $600,000-a-year pension.
- Continued health-care coverage and other benefits
- Stock options worth $8 million.
- Another $5 million worth of Time Warner stock.
- An office, secretary, and $6 million life-insurance policy for two years after his firing.
- $250,000 a year through 1999 as an "employee," even though there is little or nothing for him to do.

King of the executive suite Golden Parachutes so far, though, is Michael Ovitz, the number two Mouseketeer at Disney. Recruited by Michael Eisner, Ovitz spent about a year on the job before the two Mikies agreed it was not working between them. By many accounts the brevity of Ovitz's tenure was exceeded only by its failure—they say he was lousy at the job. Instead of a kick in the ass, though, Eisner dipped into corporate funds and gave Ovitz a payout of about $130 million for his trouble. The joke going around is that at Disney headquarters, instead of taping a "Please kick me" sign on the backs of unsuspecting colleagues as a joke, company vice presidents now tape "Please fire me" signs to their own backs.

Wait a minute—don't corporations have governing bodies, boards of directors, to monitor such self-indulgence and bring these CEOs down to earth? Yes indeedy, that is the theory, but guess who mostly sit on corporate boards? Other CEOs. And most of them are old school chums or golfing buddies of the executive whose excesses they are supposed to oversee. When John Teets of Dial Corporation had that $10 million of corporate funds put into his baseball venture, for example, he did so with the blessing of his board of directors—nine out of eleven of whom were hand-picked by him.

If chumminess alone won't make a board compliant, cash might, and plenty of it is being tossed around in the form of board fees and other freebies. These payments work like hushpuppies, the little fried balls of cornmeal, sugar, and onion served with catfish and other Southern delicacies, first concocted by cooks on hunting trips to be tossed to the dogs to "hush up" their yelping. Today a dough much richer than cornmeal is routinely fed to corporate board members—in a 1996 survey of the 200 largest industrial corporations, *Forbes* magazine found that these "watchdogs" averaged *$700 an hour* in board fees. That will buy a lot of hushpuppies! At IBM the fees amount to $91,000 a year for each member; at General Electric they get $132,000 each; and Compaq computer pays $274,000 a year. Just for a few days' work. Plus, 70 percent of the companies provide lavish pensions for their directors, worth up to $100,000 a year. If that still isn't enough to keep a watchdog fat and sleepy, companies also toss out other tasty tidbits to stop any potential yelping—General Motors directors, for example, get a new Cadillac every three months, while directors of United Airlines can fly anywhere, anytime—first-class, for free.

To bring some measure of accountability to CEO pay, shareholder groups recently have demanded that internal "compensation committees" be established to review performance and dispense the rewards to those at the top. No problem, say executives like Lowell "Bud" Paxson, who is the top guy at the telecommunications giant that bears his name. In 1997, he was honored to learn that his compensation committee had seen fit to award him a generous bonus of $1,875,000. Guess who was on the committee? Paxson himself! Who else served? No one—he was the sole member. Talk about cronyism. Can't you just imagine that committee meeting? Bud: "I move we give Bud a fat bonus." Bud: "I second that." Bud: "All in favor say aye."

Bud: "Aye." Bud: "Let the record show the vote was unanimous." Bud: "Drinks for everyone in the house!" Not bad work, if you can get it.

Curiously, both these executive suite excesses and the insistence on harsh austerity for everyone else in the company are brought to us under the same banner: "global competitiveness" (Translation: Japan made me do it). Never mind that Japan, Germany, France, and some of our other major competitors in the "new global economy" pay higher wages to working people, provide far better benefits, and generally treat workers with much more respect and sense of partnership than American CEOs (thereby rendering specious the corporate argument that knocking down our country's working class is a competitive necessity), let's deal here with that self-serving notion that "too much is not enough."

The last time I checked, Japanese executives were still beating the bejeezus out of our overstuffed CEOs, and they were doing it without being fed a number six washtub full of cash each morning. The highest-paid Japanese executive is Hiroshi Yamauchi, head of the Nintendo company. He makes just over $6 million a year—plenty enough to put sushi on the table every night, but nowhere near the embarrassment of riches heaped on American executives. Only six Japanese CEOs take more than a million dollars. Honda's top man, for example, gets $632,000; Cannon camera's CEO gets $584,000; Kobe Steel's gets $436,000; and Japan Air Lines' CEO gets $391,000. Remember that ratio of U.S. execs typically getting 225 times the average pay of a factory worker? In Japan, the top pay is thirty-two times greater.

Now here is something that will make you laugh till you cry: If a Japanese corporation has a poor year, the CEO takes a pay cut! Imagine! Not a little symbolic slap on the wallet, either—more like 35 percent. And (grab your socks now, you won't believe this) if it is a really bad year, like maybe so bad that the company has to let some workers go, the CEO *resigns*. "When company results are down," a Hitachi honcho told *Business Week*, "managers have to take responsibility." Imagine. Maybe it's all that tofu they eat.

Years ago chewing gum baron Philip Wrigley said: "There are highly successful businesses in the United States. There are also highly paid executives. The secret is not to intermingle the two." I think he was on to something there, so here is a proposal to help get the American economy back on the right track: Instead of continuing to let these corporate executives export our jobs and futures to Third World countries that pay between a nickel an hour and 50 cents an hour for labor, while the same executives fatten themselves like a hog on steroids, why don't we *import* some of those half-million-dollar-a-year CEOs from around the world who have shown they can run successful companies without running over their own employees and destroying the middle class? If American industry must become "lean and mean" to compete in the twenty-first century, let's start at the top.

RUTH'S BRICKS

Let me introduce you to Ruth Shaver of Mesquite, Texas.

She worked for Safeway for twenty-two years until one day in July 1986, when something called "Kohlberg Kravis Roberts" dropped out of the sky and took over her company . . . and her job.

KKR is a firm of Wall Street pirates that specializes in raiding corporations. Armed with more than $4 billion in junk bonds, they leaped on Safeway and claimed it as their own—three gentlemen who would not know how to bag a line of groceries if their lives depended on it suddenly owned the country's biggest supermarket chain.

Four billion dollars is a hell of a debt load to carry, even for pirates, so to lighten it, KKR immediately began ripping up big pieces of Safeway's assets and throwing them overboard. Believe it or not, this is what passed for good business sense in the eighties. Among the assets they chunked was the entire Dallas division of the company—141 Safeway stores, along with 8,814 Safeway people. Bam! Out the door.

Ruth Shaver was one of them.

No one was a more cheerful, dedicated, loyal, and capable employee than Ruth was for Safeway. She checked groceries, bagged them, stocked shelves, trained others, and helped remodel stores. She took great pride in her work and had the kind of attitude that caused customers to ask for her by name, and if she was checking, they'd wait in her line even if it was longer than the others. For a lot of people, Ruth Shaver *was* Safeway.

Over the years, she worked her way up to wages of $12.06 an hour, about $24,000 a year—not enough to summer in France, but enough to buy a small slice of middle-class America.

With no notice, though, or fare-thee-well, her store was abruptly closed and she was dumped. "They left me high and dry," Ruth says. Because of her union's contract, she did get a small severance payment, but it lasted only a few weeks, after which she was stretched to the breaking point. Of course, she had begun looking for work immediately, but she was one of 8,800 people put on the market at once, and the Dallas job market in the late eighties was already a disaster.

It took her seven months to land another job—not one at twelve bucks an hour, but $5.70, plummeting her from that middle-class income of $24,000 to under $12,000 a year. At these wages, she couldn't keep up the payments on her Buick, so she lost it, along with her good credit rating.

But Ruth is a scrapper. She got an old clunker of a car and began struggling back. After a couple of years, she was working two jobs—she would leave her trailer house at 10:30 P.M. for the night shift at a warehouse, get off at 7 A.M., then drive across town for an 8-to-noon part-time position at a local grocery store. Ten years later, she is finally working full time again at a grocery store, and is up to $10.85 an hour. Also, she now drives a '90 Hyundai, though she had to buy it in her son's name since her credit was shot, but even that is beginning to recover—she recently got a Penney's charge card.

Like Ruth, the 8,813 other Safeway employees were a long time finding other jobs (a third were unable to find employment for more than two years), and when they did get jobs, practically everyone had to swallow pay cuts of 30 to 60 percent. Four committed suicide. The divorce rate among these families has been epidemic, and the former employees have suffered an abnormally high number of heart attacks.

On the other end of the Safeway takeover, though, the picture has been a whole lot brighter. Jerome Kohlberg, Henry Kravis, and George Roberts each enjoyed personal incomes of more than $40 million the year they bounced Ruth and her coworkers. Henry Kravis has become one of the 400 richest people in America and often gets his picture in the society pages partying with Oscar de la Renta and the high-hat fun crowd in New York City. He was cochair of New York's "Bush for President" finance committee in '88. But don't pigeonhole him as an unbending Republican—in that same year, he was a major contributor to Senators Bill Bradley, Lloyd Bentsen, Daniel Patrick Moynihan, and other Democratic lawmakers up for reelection. (Call me a galloping Trotskyite if you must, but am I the only one who suspects this financial cross-pollination by the Kravises of our country is the reason neither party ever does anything about their piratical ways?)

Then Mr. Kravis took the plunge that so many superfluously wealthy corporate marauders take: He became a "philanthropist." Not a modest one, of course—Henry wanted full credit and a major public fuss made over him, and you can be assured he grabbed a full tax deduction for his generosity. He donated $10 million to New York's Metropolitan Museum of Art. In return, the museum named a wing of its building after him.

So Henry Kravis got the glory as well as the gold. But, you know, there are bricks in that wing of the Metropolitan Museum of Art that belong to 8,814 Safeway employees, including Ruth Shaver, who still lives in a trailer house in Mesquite, Texas.

ST. PETER

There must be a Republican gene.

How else can we explain the astonishing fact that Republican politicians from Warren Harding to Ronald Reagan and on now to Newt Gingrich have been so consistently committed to the preposterous notion that the answer to poverty is charity and that the only way to increase charity is first to increase the fortunes of the wealthy. "If ye truly want to help the poor," goes the dogma of Republicanism, "first provide tax breaks, subsidies, and other incentives to spur the blessed rich to do what they do best (which is to amass greater and greater wealth unto themselves), then put ye your trust in the charitable nature of these good and wealthy souls."

There are at least a couple of snakes under that log. First, every time we have tried this trickle-down philosophy, we have indeed made the rich fabulously richer, but somehow or other, the poor have always ended up getting trickled on, figuratively speaking. Under Reagan, for example, millionaires made out like bandits, but their charitable natures did not expand, they shriveled. Not since the Roaring Twenties had the incomes of the rich risen so dramatically, yet the level of contributions by millionaires to charities during the eighties fell from more than 7 percent of their income to less than 4 percent, and precious little of that went to poor folks. The preponderance of charity from elites goes to universities, museums, churches, and other prestigious and powerful institutions.

The second and more fundamental flaw in the "let them eat charity" philosophy is that it is a miserable and miserly substitute for developing a real economic policy that gives people the tools they need to build and maintain middle-class life—basic tools like stronger labor laws, health care for all, true educational opportunity, affordable housing . . .

There is a third flaw, too, and it really sticks in my craw. Too much of the "charity" bestowed by the wealthy comes from those who are not really charitable at all. They have built fortunes on the backs of people to whom they pay poverty wages, then they want public applause for donating a fraction of a percent of that ill-gotten fortune to the poor—like the spectacle of John D. Rockefeller himself in his later years, going around doling out dimes to children.

I'll tell you what St. Peter thinks about trickle-down. This is a story that old Earl Long used to relish telling when he was governor of Louisiana:

> A rich man died and tried to get into heaven. As you know, though, one doesn't just waltz through the Pearly Gates, no matter who you are. No, no—you must present yourself to the Angel standing outside, and that Angel reviews your life while St. Peter listens, then St. Peter renders judgment on whether you get to enter the Kingdom of Heaven.
>
> So the Angel looked over this rich man's life, shook his head, and said: "Why, you never did no good for nobody no way; why are you even standing here?"
>
> "Now wait a minute," the rich man said, "I have given to charity. Why, in 1933 I came across a poor widow who needed food and I gave her a nickel."
>
> "So?" the Angel asked.
>
> "Well, then there was that time in 1943, I saw a beggar man on the street, and I put a nickel in his cup."
>
> "Maybe so," the Angel said, "but that doesn't make up for a lifetime of greed."
>
> "But don't you see," the rich man pleaded, "I have a consistent pattern of philanthropy. In 1953, filled with the Christmas spirit, I came out of my bank, walked right over to the Salvation Army kettle, and put a nickel in there."
>
> The Angel turned around to St. Peter, threw open his arms, and asked, "What in God's name are we going to do with this man?"
>
> St. Peter scowled and said in his thunderous voice: "Give him back his 15 cents and tell him to go to hell."

A SHINING CITY ON THE HILL

I still have a glorious memory of my earliest encounter with a merry-go-round, way back when I was but a tyke. This particular carousel was a part of a small penny arcade and amusement park at Burn's Run Beach on Lake Texoma, near my hometown of Denison. I'm sure now it wasn't much of a whirligig, as merry-go-rounds go, but you couldn't have told me that then. I was dazzled by this marvelous spinning contraption, its glittering lights, its blaring calliope music, its swirling horses, camels, and other critters—even a yellow and green seahorse! My eyes grew larger than 50-cent pieces, my heart drummed faster than the music, and my entire life's ambition at that moment was nothing less than to get on board and ride, ride, ride.

Good old American capitalism strikes me in much the same way today, pulling us—as well as so many immigrants from around the world—toward its irresistible sparkle and razzmatazz. Oh for a ticket to ride on such a fantastic, twinkling system!

There, though, is the sticking point. One of the least acknowledged oddities about our capitalist society is that so very, very, very few of us are. Capitalists, that is.

I do not mean ideologically, but tangibly. How about you? Do you have much capital to speak of—capital being the wealth to invest so that your investments might spin out more wealth for you to invest, which would spin out still more . . . and on and on, round and round, spinning faster and faster on that phantasmagoric merry-go-round?

If you are like most of us, what you have is your take-home pay, a house (mortgaged), a car or possibly two, a life-insurance policy, maybe a small pension or savings account, a closetful of clothes, a shelf of good books, a full set of Tupperware, your bowling ball, some socket wrenches, and the complete recordings of Engelbert Humperdinck. This does not a capitalist make. It would not take much more than a good yard sale to cash you out, and even then you would barely cover your debts, much less be in a position to do lunch with Donnie Trump.

Warning: Statistics Ahead, Next Three Paragraphs. Us middle-classers mostly own no stocks and bonds, investment properties, businesses, or trusts. More than two-thirds of our "wealth" is tied up in our homes, with about 20 percent more in life insurance, checking accounts, savings deposits, and other small-potato holdings. For us, wealth is our safety net, a little something to fall back on, rather than a power lever to rocket us upward to higher and higher levels of . . . well, of capitalism.

This is what makes us different from the superrich—the wealthiest 1 percent of American families who are worth more than $2.3 million each. These fortunate ones put less than 7 percent of their wealth into their homes (which still gives them houses with more square footage in linen closets than our houses have in bedrooms); instead, more than 80 percent of their savings go into stocks, real estate, and other investments that pack a powerful financial punch and steadily return big bucks to them. *These* are the genuine capitalists. There are only some 600,000 of these families in our entire country, but they own 46 percent of all the stock in America, 54 percent of all the bonds, 53 percent of all the trust accounts, 56 percent of all the business equity, and 40 percent of all the investment real estate.

Add to this wealthiest 1 percent of families the next-wealthiest 9 percent (those whose net worths are $346,000 to $2.3 million each)—and you have pretty well lassoed all of our country's practicing capitalists. This top 10 percent of our population owns 89 percent of the stocks, 88 percent of the bonds, 89 percent of the trusts, 90 percent of the business equity, and 80 percent of the investment real estate. These privileged few have attained this astonishing level of wealth concentration the old-fashioned way: They got our government to give it to them.

If poverty is its own punishment, wealth is its own reward, bringing not only comfort, social status, and raw economic power, but also tremendous political advantage, including the ability to buy politicians (OK, maybe they don't actually buy them, but they damn straight take out long-term leases on them).

When Ronald Reagan was President, he described America as "a shining city on a hill." What an elegant phrase! But what he didn't mention was that you and I—the workaday majority—would not be living in that shining city; no indeed, our place was well down the side of the hill toiling in the shadows, doing the drudge work to sustain the sumptuous lifestyles of those at the top. The few have always had the most in our country, but in the past two decades, Reagan, then George Bush, and now Bill Clinton—along with loyal, cheering congressional majorities forged among both political parties—have teamed with America's most privileged and powerful citizens to create an unprecedented rise in this wealth inequality. They have done it by gaily dismantling our nation's progressive tax policies, contorting our country's regulatory policies, rigidifying and channelizing monetary policy, diverting our spending policies, and perverting America's trade policies. All of their reengineering and deconstructing has resulted in massive, unprecedented loads of cash and economic power being hauled out of our neighborhoods up to the top of the hill.

One final load of statistics: In the eighties and nineties, *practically all* of the new economic wealth that we generated as a people—some $12 *trillion* worth of gain—went into the pockets and portfolios of the wealthiest 20 percent of us. Indeed, *nearly two-thirds* of the overall increase in the American people's net worth went to that handful of superprivileged citizens, the wealthiest 1 percent.

This is such a phenomenal fact, such a staggering accumulation of wealth in so few hands, that it absolutely begs for a bar graph. I have done my best to avoid imposing such wonkishness on you, and I apologize in advance, but I simply cannot help myself this time. Here goes (sorry):

SHARE OF AMERICA'S GAIN IN WEALTH, 1983–89, BY WEALTH CLASS

RICHEST 1%	61.6%
NEXT RICHEST 9%	28.7%
NEXT RICHEST 10%	8.5%
YOU (BOTTOM 80%)	1.2%

Source: Edward N. Wolff, New York University, editor of *Review of Income and Wealth*

Professor Edward Wolff, whose recent book *Top Heavy* documents the stunning rise of America's wealth inequality, says that figures since 1989 indicate that the top 1 percent are now taking 68 percent of the total increase in household wealth, so "the superrich are moving away from the rest of the population at an even faster pace in the '90s."

During the Reagan regime, the Reaganauts referred to their policies as "supply side economics." Now we know what that means: Their side gets all the supplies. Just twenty years ago the 600,000 superrich families held 19 percent of America's total wealth—add up the value of every household's cash, stocks and bonds, cars, liquor cabinet, lava lamps, VCRs, everything . . . and they owned 19 percent of it. Fast forward to today. They now own more than 42 percent of everything.

THE NEXT BEST THING TO SLAVES

At last U.S. industry has figured out how to compete with Third World wages right here at home. Hire prisoners! No need to mess with the want ads, employment agencies, or job fairs to find cheap workers, just bustle on down to your state prison and cut a deal for some convicts. Since 1990 thirty states have contracted out prison labor to private companies.

- JCPenney, Kmart, and Eddie Bauer are getting such products as jeans, sweatshirts, and toys made by prisoners in Tennessee and Washington State.
- IBM, Texas Instruments, and Dell computer all get circuit boards made by Texas prisoners.
- Honda has had car parts made in Ohio prisons, McDonald's has uniforms made in Oregon prisons, AT&T has hired telemarketers in Colorado prisons, and Spalding gets golf balls packed in Hawaii prisons.
- California's correctional system has become a one-stop hiring hall for corporations: San Quentin inmates do data entry for Chevron, Macy's, and BankAmerica; Ventura inmates take telephone reservations for TWA (yes, this does mean callers are unwittingly giving their credit card numbers to criminals, and, yes, there have been "incidents"); Folsom inmates work for both a plastics manufacturer and a brass faucet maker; and Aveala inmates run an ostrich-slaughtering facility for an exporter that ships the meat to Europe.

Who says American industry is losing its ingenuity? These free-enterprisers not only get labor for minimum wage and less from the state, but they also provide no health care, no pensions, no vacations, none of those other frills that pampered softies on the outside are always crying about. Plus these jailbirds always show up on time for work, they don't call in "sick" to go to a ballgame, if they talk back you can have 'em thrown in solitary, and they darn sure won't be joining some pesky union. I tell you, it's the next best thing to having slaves—maybe better, since the company doesn't even have to feed and house them.

Oh, and here's the best part of all: You can slap a made-in-the-U.S.A. label on every product they make for you!

Convict-made goods are expected to reach nearly $9 billion in sales by the end of the decade as the prison population swells, as more companies discover the scam, and as more state politicians learn to cash in on it. Wisconsin governor Tommy Thompson, never one to pass up a chance to exploit someone's misery, has been especially adept at huckstering his state's prison force: "Can't find workers?" a state mailing asks corporate executives all across the country. No problem, proclaims the brochure, "A willing workforce waits"—conveniently incarcerated for you in Wisconsin.

Most companies pay the minimum wage, but many get away with paying far less—AT&T, for example, paid only two bucks an hour for its imprisoned telemarketers, and Honda got its convict-made car parts from the Ohio prison at two-and-a-nickel an hour. The prisoners typically get to keep only 20 percent of the paycheck, with the state government grabbing the rest, which is why the states are all for it.

Participating firms everywhere sing the praises of this locked-up labor. In an article in *Nation* magazine, Bob Tessler of DPAS company in San Francisco gushes: "We have a captive labor force, a group of men who are dedicated, who want to work. That makes the whole business profitable." That, plus the fact that California taxpayers also give Tessler a 10 percent tax credit on the first $2,000 of each inmate's wages. Wow, cheap prison labor *and* a subsidy—if that won't restore your faith in the working of the free market, nothing will! It is such a steal of a deal that Tessler has shut down his operation in Mexico, moving his data processing work inside San Quentin. "Here we don't have a problem with language, we have better control of our work and, because it's local, we have a quicker turnaround time."

LABOR DAY

There is at least one of everything in New York City, so it was no surprise at all to encounter a long-haired, thirty-something fellow in jeans and T-shirt wending his way silently down Broadway through the morning rush-to-work crowd, holding aloft a hand-lettered message on a big piece of cardboard. Still, I noticed that even jaded New Yorkers winced at the punch this guy's cardboard threw at them as they hurried to the office:

Consume
Watch TV
Be Silent
Work
Die

Ouch. But rather than wince, why not pick up on number three? We will continue to consume, watch TV, work, and die—but we do not have to keep silent as the powers-that-be cut the incomes and security of the workaday majority, cut back on our health care and pensions, cut corners on job safety, and ultimately cut out on the middle class altogether.

There is no need for this class war. America is not suffering—we are in high cotton! Our economy is spinning out more wealth than ever; worker productivity is the highest in the world; corporate profits and stock prices are at *astonishingly* high levels; executive pay is an embarrassment of riches. Yet 80 percent of us are barely getting along and seeing more and more of our opportunities closed off.

Even a dog knows the difference between being stumbled over and being kicked.

What America's economic elites really have severed is the essential American idea of all-for-one-and-one-for-all—that if we work hard, if we are loyal and creative, if we help raise the tide, we'll share in the gain. This is the "social contract" that holds us together, not merely as an economy, but as a society.

Bear in mind that this contract was not "awarded" to us—we

demanded it, fought for it, bled and died for it . . . and won it. The forty-hour week, overtime pay, the very idea of a minimum wage, collective bargaining, social security, Medicare and Medicaid, pensions, and all the rest that so many families have taken for granted for so long came against the wishes and the bitterest opposition of the powerful.

Even Labor Day. Schools do not teach it and history books ignore it, but even getting a simple holiday for workers required an epic struggle. For most people, Labor Day is just a time to hit the pool or the mall, fire up the grill for an evening cookout, do some twelve-ounce elbow bends through the day to stay limber, and maybe toast the bosses for getting the day off.

Toast the bosses?! Holy Joe Hill, the bosses did not give you the day off! Hell, they damn near choked on their fine whiskey and swallowed their big cigars at the mere mention that there should be a holiday celebrating the principle that *workers* create wealth.

To the contrary, Labor Day was a bottom-up creation, our first national holiday put on the calendar by ordinary people. Dauntless Matt McGuire of the New York City carpenters union was the one who first proposed that on September 5, 1882, all labor unions in the vicinity should make "a public show of organized strength." This was no bid for a day at the beach, either. People were working twelve-hour days, six days a week, for a daily wage of $2, and this parade of unionists was the shot across the bow of the Gilded Age tycoons, declaring an all-out fight for an eight-hour day at fair pay. In support of McGuire's proposal, the New York Central Labor Committee unilaterally declared September 5, 1882, to be a workers' holiday, and called on working people to march right through the heart of the city and on into the battle for fairness: "We are entering a contest to recover the rights of the workingmen and secure henceforth to the producer the fruits of his industry," trumpeted their official call to arms.

And they came. Defying bosses and risking both their jobs and their personal safety, thousands of bricklayers, longshoremen, jewelers, carpenters, typographers, cigarmakers, and others marched in rank and file with bands blaring, banners streaming, and fierce pride proclaimed in their every step. From City Hall up Broadway to

Union Square they proceeded, 4,000 strong, then across Seventeenth Street to Fifth Avenue, where they turned north. This was a sight hard to imagine: row after row after row of $2-a-day laborers, six abreast, tramping straight up what was then the most ostentatious corridor of wealth and power in America. Reporter Richard Hunt described the scene: "They passed August Belmont's house; they trudged on past the tonish Brunswick Hotel; past the uptown Delmonico restaurant; past the elegant new Union League Club; past the mansion of Vincent Astor. Mrs. Astor—along with many of her millionaire neighbors—was in Newport for the season. Nonetheless, if the consciousness of capitalism was not penetrated, its precinct was."

The spirit of the day was so powerful and the march was such a rip-snorting success that New York's labor committee resolved to observe the first Monday of every September as "Labor Day." Bosses railed against this usurpation of authority and forbade it, editorialists declaimed it as ruffians on parade, politicians denounced it as anarchy and rank ingratitude to employers. Still, the workers took off. Within a dozen years, workers in twenty-five states had appropriated this same day as their own, using it as a symbol of their determination to gain a better share of their own productivity. By 1894 the movement's strength was such that Labor Day was sanctioned by an act of Congress and signed into law by Grover Cleveland. The people had taken their own day off.

A hundred years later, the powers-that-be again want to deny us the gains of our productivity. They have amassed so much of our society's money and power that they now feel free to decouple their fortunes from the well-being of the rest of us, kicking down the whole structure of middle-class opportunities we Americans have built up since the days of the robber barons—a structure that our very society is built on.

In its place they supplant our hard-won social contract with a corporate ethic of "I got mine, you get yours," "Never give a sucker an even break," "Caveat emptor," "I'm rich and you're not" . . . "Adios, chump." Even more perniciously they tell us that we don't matter in their newer, braver economic world. By word and deed they say:

> "We can make you compete with Third World wages, reduce you to temporary and part-time status, force your family to work two or three jobs to make less than one wage earner used to bring home, and treat you as disposable—because you don't matter."

> "We can deny you health coverage, loot your pension, price both college education and home ownership beyond the reach of you and your children, and make your whole family downwardly mobile—because you don't matter."

> "We can devalue the minimum wage, compel you to work sixty-hour weeks, eliminate overtime pay, make a mockery of collective bargaining, and even take back that Labor Day if we choose—because you don't matter."

The "you" includes not only you the low-skilled worker, but also you the highly skilled machinist, the college-educated computer programmer, the teacher, the professional engineer, the architect, the mid-level executive . . . *you*, the entire middle class.

It is said that the rich and the poor will always be among us. Perhaps so, but it is not assured anywhere or in any time that the middle class will always be among us. Quite the contrary, a middle class is created only through great struggle by the people themselves, organizing, communicating, connecting, rallying, demanding, striking, marching, fighting, agitating, and undertaking all the other untidy activities that have been required throughout our history for ordinary people to make gains—and to hold them.

The powers-that-be do not want us thinking like this. Be silent! But as happens periodically in our country (1830s, 1880s, 1920s), the rich are not merely getting richer in the 1990s, they are getting ridiculous, and we have to fight for our economic freedom. This will not be done *for* us, only by us.

From time to time, I have been jumped by apologists for the status quo who get right in my face, shouting, spewing spit, and demanding to know: "What is it you want anyway?" Fair question. I want what working people have long wanted:

More schoolhouses and fewer jails.
More books and fewer arsenals.
More constant work and less crime.
More leisure and less greed.
More justice and less revenge.
In fact, more of the opportunities to cultivate our better natures.

This list is from a speech by Samuel Gompers, founding president of the American Federation of Labor, in 1893.

EARRINGS ON A HOG

My friend Verne Newton once attended the official lighting of the National Christmas Tree. Each year, several thousand people in our nation's capital bundle up and trek over to the Ellipse, the small park immediately south of the White House, where they stand in the cold until the tree's lights are flicked on, then they leave. Washingtonians are fairly easily amused.

Verne reports that on the night he attended there was a jolly crowd waiting expectantly in the December darkness, all eyes trained on the shadowy hulk of what he took to be at least an eighty-foot-tall, eighty-year-old ponderosa pine hacked down and hauled in all the way from someplace like Extremely Tall Trees, Montana. Next to the towering pine was a small platform on which the President, the First Lady, the First Dog, and a whole mess of picture-perfect children collected. The crowd had been waiting for more than an hour, but at last everything was about ready for the Leader of the Free World to press the button that would suddenly emblazon every branch of this noble ponderosa with electric Christmas color.

As the joyous moment approached, Verne thought he heard someone behind him call his name. Reflexively he turned his head, but he recognized no one, and he just as quickly turned back for the lighting of the tree. Too late! In that instant of turning, the button had been pushed and the tree was lit. Verne was there, but he missed it.

It occurs to me that a lot of conservatives must have had a "Verne experience" when it comes to observing the workings of our economy. Say to them, as I have in my political travels and talk-radio conversations, that the rich are getting richer and the rest of us are getting taken, and quite often they'll curl their lip and snap back with this verbal sneer: "Why, you're just trying to start a class war."

Too late for that, Limbaugh-breath! Maybe you missed it, but that button was pushed a long time ago.

You want class warfare? Check out how corporate lobbyists have conspired with their hired hands in Washington to hold down the minimum wage, which exists as our society's income floor, a base level of pay that says, "This is how we value people who work." Forget the PR crap put out by low-wage employers that the only Americans toiling for the minimum are a few suburban teens grabbing extra spending money by working at the Whataburger after school. Not so. Eleven million Americans are employed at this bottom-level pay, and nearly two-thirds of them are adults, most trying to support families.

Just as important, the wage structure for millions of others is built on this floor. Raise the floor, and those making one, two, or three bucks above the minimum end up benefiting, too, making the minimum wage one of the best tools we have for spreading the gains of worker productivity.

Candidate Clinton pledged boldly and unequivocally in his 1992 run for the presidency to lift the $4.25-an-hour floor that holds so many families in peonage, asserting that poverty ought not to be the reward for work in our fabulously wealthy economy. Good for him!

Unfortunately Clinton operates like a light bulb not screwed into its socket very tightly; he occasionally flickers with brightness, but you never get any steady light. No sooner had he settled into the big chair with the presidential seal on it than he abandoned those families. Yeah, yeah, he did make mention of raising the wage in his first State of the Union message, but it was all a pose—he never even sent a bill up to Capitol Hill, despite the fact that his party controlled the Congress and that 75 percent of the public was in favor of a major minimum-wage boost.

What spooked Bill? Wall Street. Goldman Sachs, Bear Stearns, and the other big bond houses hold way more sway in Clinton's White House than do the houses of the working poor. These are generous contributors to Clinton, they have allies like Treasury Secretary Robert Rubin positioned inside the administration, they have lobbyists like Vernon Jordan who are close pals and golfing partners with the Prez, they have key congressional leaders in their pockets—and all of these voices gave the same single word of advice to Clinton in 1993: "jittery." Major Wall Street investors, they advised the President with all the solemnity they could muster while saying something so patently silly, would be "jittery" at any effort by him to lift wages, and the very mention of it could precipitate (gasp!) a decline in the Dow

Jones Average. Showing the same steely-eyed firmness that would characterize the rest of his presidency, Bill said, "Oh, OK."

And that was that, until three years later, when his reelection campaign was under way. In the spring of '96, responding to rising public anger that working folks were being trampled by unbridled corporate greed, Bill decided to toss the dogs a pork chop, and suddenly Candidate Clinton was back, talking once more about the moral imperative of raising the minimum wage "to make work pay." But not much, mind you, nowhere near a living wage and not even enough to lift a full-time worker out of poverty. With much fanfare and garlands of self-congratulation, he and the Republican Congress approved a boost of—*ta-ta-ta-la, ta-la, ta-la*: 90 cents. And even this parsimony has been phased in over fifteen months, spending a year at $4.75 an hour before finally reaching its full flowering of $5.15 an hour. Clinton privately told a couple of labor leaders that he would like to have done more, but he did not want to risk making Wall Street . . . well, jittery.

Wall Street can relax. Clinton's big 90-cent increase still leaves minimum wage workers getting less in buying power than they got in 1956, when Ike was President. Try buying a slice of the American dream on Clinton's minimum wage, which pays about $860 a month gross for full-time work:

A Month on Clinton's Minimum-Wage Budget

Rent:	$350	for even a rundown nothing-of-a-place
Utilities:	$50	just for gas and electric, no phone
Household:	$200	to put groceries on the table, buy toilet paper, soap, sheets and towels, cleansers, and other essentials
Transportation:	$50	to get back and forth to work, assuming a bus or subway even goes there; way more if you need a car; and this does not count any errands, visits, etc.
Personals:	$50	for clothing (including work uniforms, which employees have to buy or rent), haircuts, cold medicines, etc.

Taxes:	$50	just for social security, not counting other federal, state, and local taxes, or even fees for garbage pickup, etc.
Subtotal:	$750	

This leaves $110 for the rest: child care, medical care (don't dare get really sick, because you don't have any insurance, and emergency rooms cost a week's wages just to talk to a nurse), education or job training, amusements (no cable TV, no movies, no ballgames, not much of nothing), a home computer for your child, vacations (ha!) . . . and what about Christmas and your child's birthday? It makes you ache, inside as well as out. As one lady put it, working for minimum wage—even after Bill's Big Boost—"leaves you feeling like you've been ironing all day on a low board with a cold iron."

In his January 1998 State of the Union message, Ol' Bill waxed rhetorical once again about raising the minimum wage, giving it a one-sentence plug as "a simple, sensible step to help millions of workers struggling to provide for their families." Good line, but once again, weak follow-through. His proposal was to raise the wage by only another buck. Even if it happens (and Newt Gingrich, Trent Lott, and their Republican majority pointedly sat on their hands and glared stony-faced at Clinton as he delivered his rhetorical gesture), it would mean that someone working full time still would be mired in poverty, grossing just over $12,000 a year, minus payroll taxes and other deductions. Six years into his presidency, at the height of his popularity, and Clinton still would not spend a couple of political chips to honor his 1992 campaign promise that the reward for work would at least be higher than the poverty line.

You want class warfare? Washington's corporate lobbyists are not nicknamed "cockroaches" for nothing—they swarm in the dark, feed on anything that hits the floor, and consider no act too low.

While Clinton and the Congress were passing their miserly minimum-wage hike last year, the roaches were climbing on the legislation like it was birthday cake, scurrying off in the dark with some $21 *billion* worth of corporate gimmies from a bill ostensibly intended to help the working poor:

- A "clarification" of tax law was tucked into the bill, allowing newspaper conglomerates to classify their carriers—the minimum-wage folks who deliver your morning paper—as "independent contractors." This allows the Murdochs, Coxes, Dealeys, Hearsts, and other billionaire publishers to avoid paying social security, unemployment, and other benefits to those minions whom they so hypocritically hail in an annual editorial on Newspaper Carriers Day.
- U.S. multinational corporations snuck in a loophole eliminating taxes on income they make from their foreign factories, thus giving them yet another tax incentive to move more of our factories and jobs abroad.
- Corporate raiders like Henry Kravis, Ron Perelman, and others who conduct hostile takeovers of corporations, then fire the employees and plunder the assets of the takeover targets, got a great big goodie from the minimum-wage bill, too. These pirates pay billions of dollars in fees to investment bankers to finance their job-destroying raids, but now—thanks to a "technical correction" jammed into the wage bill—those banking fees will be tax-deductible. Even more fun, Congress made the deductibility *retroactive.*

At the White House signing ceremony for the minimum-wage bill, Clinton surrounded himself with low-wage workers for a photo op, but off-camera, it was the Wall Street powerhouses who had the broadest grins.

You want class warfare? It's bad enough that top corporate executives hold down everyone's wages and fire working people right and left while they scamper over the hill with $5-million-a-year, $50-million-a-year, $150-million-a-year paychecks in their pockets, but to make the deal even rawer, Washington forces you and me to subsidize the salaries of these profligates. No matter how extravagant, every dime of the CEO's salary, bonus, extra compensation, and perks is deducted from the corporation's tax bill.

If you pay your teenagers to mow your yard twice a month, do you get to deduct this expense from your taxes? No, you do not. If you did, you undoubtedly would pay them $100 or $1 million, or as much as you could for each mowing because, well, why not . . . the taxpayers are covering it for you.

Once again, Bill Clinton's light flickered on and he expressed his determination to do something about runaway greed in the executive class. In his first State of the Union speech, the fresh young President stood tall at the national podium and dared to suggest that, as an egalitarian gesture, he would restrict a corporation's tax write-off to "only" the first million dollars of a CEO's swollen salary.

"Great God Almighty," every chief executive shrieked in unison from top-floor suites all across America, "fetch a rope and call the hounds, there's a Bolshevik loose in the White House!" Even before Clinton completed his speech, the phone calls were made, lobbyists were hired, and the "Save the Millionaires" campaign was launched to try to stop this act of "cheap populism," as they quickly branded the new President's proposal to tighten their loophole.

They took this frontal assault on their class privilege very personally, with top executives themselves getting involved in the fight for the preservation of fat paychecks. As reported in the *Washington Post,* the serious squawking took place in some of those out-of-the-way places where the concentric social circles of the Washington and Wall Street elites overlap. Scene 1: A black-tie dinner at a Florida resort, where some of the nation's biggest industrialists flock to spend February, fleeing the northern winter. Letitia Chambers, a White House aide, is cornered by a covey of the well-heeled ones: "Why are you picking on us?" demanded the wife of one of the well-over-a-million-dollar-a-year executives, her face flushed a bright red. Scene 2: A dinner of top business executives in New York the night following Clinton's address. The honchos of American Express, RJR Nabisco, Loews Corp., AIG insurance, and other giants are there, moaning and groaning directly into the ears of Robert Rubin, then the head of Clinton's National Economic Council and now the Treasury Secretary. These were good ears to moan into, since Rubin himself had been a $26-million-a-year man at Goldman Sachs before joining the White House—he could feel their pain. Rubin smoothed ruffled feathers and cooed soothing words: "That's not the real Bill Clinton."

Apparently not. He quickly dropped the idea, never talking publicly about it again. In April a couple of mid-level monkeywrenchers from the Treasury Department announced a "revised proposal": Instead of capping deductions at a million bucks, the administration would allow full deductibility as long as the executive's pay was attached to some vague performance goals set by the corporation's board of directors—sort of like getting a note from your mother saying it's OK for you to keep looting the treasury.

Clinton officials explained they backed off because they didn't want to "meddle improperly" in the private sector's business. Do we look like we have sucker wrappers around our heads? No one was talking about meddling with the amount a corporation pays Mr. Big—pay him a king's ransom as far as I care, but do not take it out of the taxpayers' pockets. That is the *public's* business.

The President's abrupt turnaround reminds me of the young, duded-up city slicker who sashayed into a blacksmith shop and began to poke around, claiming to know quite a bit about the cowboy life. The bemused blacksmith was hammering out horseshoes, pulling red-hot pieces of iron out of the fire, striking and bending them into shape, then tossing them into the coal ashes, still sizzling.

The city slicker stepped over, reached knowingly into the ashes, and picked one up . . . immediately dropping that finger-singeing sucker. "Hot?" the smithy asked with a smile. "Nope," the young dude said through clenched teeth, "it just don't take me long to look at a horseshoe."

You want class warfare? It raged full-tilt behind the scenes of Washington's infamous balance-the-budget imbroglio of the 104th Congress. In public view, Newt Gingrich and Bill Clinton appeared to be in total, absolute, fundamental disagreement over every dime of federal spending, going at each other like a couple of blustering schoolboys in a playground shoving match.

BILL: You're trying to kill Medicare.
NEWT: Am not. You just want to scare old people.
BILL: You're the one scaring old people, and children, too—you want to take school lunches away.
NEWT: Do not.
BILL: Do too. Besides, I can cut the budget better than you can, you blow-dried twit.
NEWT: Oh yeah? Well, I can cut it to nothing in seven years.

BILL: But you're cheating because you're using "funny numbers" to make it balance.

NEWT: It's you that's unbalanced, you hillbilly . . .

Boys, boys, hold on! While the two of them drew all the media attention to their squabbling over how much was or was not actually getting cut from Medicare, food stamps, college loans, and other people's programs, there was practically zero discussion of a more significant question: Who was *not getting cut*? At all. Not cut by Republicans, or by Democrats.

We were offered a clue just before Christmas of 1995 when a strongly worded, double-trunk newspaper ad ran all across the country in the midst of Bill and Newt's partisan tussle: "Without a Balanced Budget the Party's Over. No Matter Which Party You're In," the headline asserted. Declaring that "America must begin to live within its means," the ad sternly called on Clinton and Gingrich to set partisanship aside, suck up some political courage, and get on with making the deep and painful cuts necessary to balance the budget, even if this meant whacking popular entitlement programs like Medicare. The ad's ninety-one signers claimed to be a bipartisan bunch united by "our common concern for America's future."

There was our clue. Anytime anyone buys two-page ads in *USA Today*, the *Washington Post*, the *New York Times*, and other pricey venues to assert their selfless concern for America, do not take your eyes off them for a single second, do not even blink, because they're trying to steal something from you. As Ralph Waldo Emerson wrote about a certain visitor, "The louder he talked of his honor, the faster we counted our spoons."

Who were these ninety-one selfless patriots? Major corporate executives, every one of them. What chutzpah, what hypocrisy! Count your spoons because, from A to Z—from AT&T to Zenith—these are the very CEOs who spent the entire budget fight in the back rooms of the Capitol and the White House, secretly waging a guerrilla war against the middle class and the poor. They made certain that all of Washington's budget balancing would be done on our backs, none on theirs. They don't want us to know it (and the media do a fine job of keeping their secret), but federal spending on *corporate welfare* far exceeds spending on human welfare, and these "welfare kings" made sure that neither Bill's nor Newt's balanced budget would touch even a dime of the $200 billion or so a year Uncle Sugar ladles out to them in the form of subsidies, special tax breaks, and outright giveaways.

Giveaways like the $85 million a year you and I pony up for something called the "Market Promotion Program" of the Ag Department, paying for the overseas advertising campaigns of major agribusiness corporations: McDonald's has taken $1 million; Campbell's Soup, $1.5 million; Fruit of the Loom, $2 million; M&M candies, nearly $4 million; Pillsbury, $9 million; Gallo, $16 million; and on down the list.

Members of Congress who rant and rave about as little as a dollar's worth of waste in the school lunch program continue to shove bales of our tax dollars into this dark orifice. Remember that amusing animated ad on television of the dancing California Raisins?—cute as could be. These raisin marketers were given $3 million of our tax funds to run their clever ads in Japan and convince the delighted Japanese to buy the shriveled fruit. But they ran the ads in English.

Despite the hue and cry for America to begin living "within its means," the corporate handouts through this Market Promotion Program not only were blessed in Newt and Bill's negotiated budget deal—they were *increased* by a third.

While Gingrich and Clinton fussed loud and long about how deep to cut Medicare, they had a quiet gentlemen's agreement to keep larding the budget with corporate goodies, including those that directly subsidize class warfare. There is the Puerto Rican subsidy, for example, which encourages U.S. firms to move our jobs to the Caribbean by giving cash to companies for each Puerto Rican job they create. Mucho cash—it averages $27,000 per year for every Puerto Rican hired, which is two to three times more than the worker is paid. Payment is based not on wages, but on the profits a firm makes. (I realize this does not make sense, but I don't think it's intended to make sense.) Drug companies have found this particular goodie especially savory, and they have used it to move thousands of U.S. drug manufacturing jobs to the island—Johnson & Johnson has been given $50,000 for every job it moved to Puerto Rico under this program, Bristol Myers Squibb has received nearly $75,000, and Pfizer has been given $156,000. That is per worker, per year, costing us a total of a couple of billion bucks annually to subsidize our own job loss.

Do you think timber giants like Weyerhauser should be cutting down old-growth trees (up to 1,000 years old) in our national forests? Not only are they, but they're also getting a $5-billion-a-year subsidy from us to do it. More than half the tall timber they fell in public forests in the Pacific Northwest is shipped off to Asia as raw logs and pulp, where those nations employ their own people to turn our timber into finished wood products. Why aren't we keeping such skilled, well-paying wood-milling jobs here for our people? Neither Clinton nor Gingrich bothered to raise this question as they quietly and quickly agreed not merely to continue this $5-billion timber company rip-off, but to jack it up by another billion.

You want class warfare? The tax code is full of it, giving breaks to corporations and the well-connected, meaning they escape taxation, and more of America's tax burden falls on you and me.

What does this sentence mean to you? "For purposes of clause 1, material terms of a contract shall not be treated as contingent on the issuance of an FCC tax certificate solely because such terms provide that the sale price would, if such certificate were not issued, be increased by an amount not greater than 10 percent of the sales price otherwise provided in the contract."

It means Rupert Murdoch and the *Chicago Tribune* company get a great big tax break.

This bit of gobbledygook involves the sale of some Murdoch-owned television stations to the *Chicago Tribune*'s parent company. A bill that would have specifically *denied* tax breaks for such transactions was sailing through congress until—wait a minute!—our congressional leaders learned that the Murdoch-*Tribune* deal was cooking and that both parties wanted a tax subsidy for it. So the lawmakers who were so intent on stopping these subsidies scrambled to make a special exception for this one. The language was snuck into the bill with no public debate, no vote, no press, no fuss. Slam, bam, thank you ma'am.

According to after-the-fact news reports, this one sentence delivered a special tax break worth somewhere between $30 and $63 million (which would make it just over a million dollars per word, if you're counting) to the Australian media baron, who just happens to be Newt Gingrich's political supporter and a big-league contributor to Republican congressional campaigns. The slick legislative maneuver also would deliver a $13-million tax break on the *Tribune* company. The Senate initiator of this fifty-nine-word lube job was Carol Mosely-Braun, a Chicago Democrat. Bipartisanship at work.

Such tax favors were laced throughout both Clinton's and Gingrich's budgets, including a brand-new loophole big enough to shove a whole *Fortune* 500 corporation through it. This one affects a few hundred tax-dodging firms that formerly manipulated the tax laws so masterfully that they avoided paying any share at all of our nation's taxes, no matter how huge their profits. In the first half of the eighties, for example, AT&T piled up $25 billion in profits, but paid zero in taxes. Correction: The phone company's accountants found so many loopholes that we taxpayers actually paid AT&T $631 million in tax rebates during this period.

To stop this nonsense, Congress passed the "Corporate Minimum Tax" in 1986, requiring at least a token payment from these loophole Houdinis. But they still managed to jiggle the system and pay a far lower federal tax rate than you do—for example, in recent years, Texaco has paid under 9 percent of its multibillion-dollar income in taxes; Chase Manhattan, 1.7 percent; Texas Utilities, 1.9 percent; and Ogden Corporation, 0.1 percent. But at least they paid something. Even this dab was too much for these fine citizens, though, so while the media had us watching Bill and Newt's noisy squabble over how much to cut the school lunch program, their lobbyists went behind the lines in Congress and quietly snipped the Corporate Minimum Tax in half. This will pull some $8 billion a year in corporate tax payments out of the treasury. That would have bought a lot of school lunches.

Republicans in Congress claim this corporate tax avoidance will "free" AT&T and the rest to put that $8 billion into job-creating investments for America, but there is no requirement that they do so, or any evidence that they will. Our experience is to the contrary—they will use it to fatten their own wallets and those of their already wealthy shareholders, and they will use it to invest abroad, moving more capital and jobs out of our country, away from the reaching hands of our middle class.

When those ninety-one CEOs ran their pious newspaper ad expressing their "common concern for America's future," our nation's chief crusader for common sense, Ralph Nader, sent a personal letter to each of them. He congratulated them on their call for everyone to make budget-cutting sacrifices and gave them a chance to join in the fun. Listing ten specific corporate welfare programs that cost us taxpayers $51 billion, Nader's letter included a pledge card for them to sign. It read:

> To help balance the budget, I have identified the federal subsidies and tax expenditures that currently benefit my corporation and selected the subsidies and tax expenditures that my corporation agrees to begin to forego immediately.

How many takers do you think Ralph got? Right.

You want class warfare? Just before Christmas of '96, Treasury Secretary Robert Rubin, the most influential voice in the Clinton Administration, issued a decree banning the use of the phrase "corporate welfare" by administration officials.

These guys are like three-year-olds who believe that if they squeeze their eyes shut real tight, we can't see them. That is delightful in a child. Less so in a leader.

You want class warfare? The high society of Wilson, North Carolina, demonstrates how bone-deep-stupid classism can get. Organizing themselves into the Wilson Appearance Committee (WAC!), these leading citizens set out to combat what they consider one of the South's greatest scourges: Tacky Porch Furniture. It seems that certain of your lower classes have been giving the upwardly mobile town of Wilson an undesirable image by moving their old living room sofas or their tattered La-Z-Boys out onto their front porches, then being so presumptuous as to plop themselves thereon just pretty as you please, and commencing to wave or even speak to passersby.

To deal with this creeping couchism, WAC worked itself into a high state of ordinancy and, in the Year of Our Lord 1998, flat outlawed unauthorized porch furniture. Not that these WAC-worthies are antiporches or even anti–porch furniture—you just have to have the good taste in Wilson, please, to purchase a $1,000 wicker set or some other proper settee capable of winning approval by the WAC Porch Police.

Deborah Thompson is one of the newly outlawed porch sitters in Wilson, and she can't see what the fuss is about. She knows that her vinyl and chrome couch is not going to make *House Beautiful*, but "this is not junk," she says. Indeed, it is her connection to her community. Ms. Thompson told the *New York Times* that she doesn't have a social calendar, or even a car, but she can—or she once could—sit out on her porch and socialize. She's not bothering anyone, especially not those WACers who live across town. "I watch the cars go by, and I say 'hey.' I sit there to keep from sitting in the house. It gets lonely in the house," she says.

You want class warfare? The powers-that-be are telling America's middle-class workers they should not worry about the outflow of manufacturing jobs because their future is in the "knowledge industry"—programming new software for Microsoft, generating interactive computer technologies for Disney, and engineering new systems for IBM. But hold your horses. These same companies are presently pushing legislation in Washington that would let them bring more low-paid, foreign workers into the United States to take these very jobs.

Already, 65,000 immigrants from India, Russia, and other countries are coming into our country each year under special H–1B, high-tech worker visas to take these jobs, yet all the High Lords of High Techdom are pressing both the Clinton White House and the Republican Congress to nearly double the annual quota of H–1B visas. If you think they just pulled that number of visas out of thin air, note that the Commerce Department recently reported that our computer industry will need 138,000 programmers, engineers, analysts, and other "knowledge workers" each year for the next decade. Under the industry's bill, 125,000 of those would be filled from abroad.

Well, yes, say lobbyists for the megabillion-dollar computer giants, but there simply are not enough Americans up to snuff on high-tech skills to fill our needs. Bull stuff. Never mind that both industry and government have been pledging throughout the nineties to make sure that today's graduates are prepared for these "jobs of the future," the fact is that good ol' U.S.A. programmers, engineers, and such are available, qualified, and ready right now to take the 65,000 jobs currently being claimed by the foreign nationals. But here's the rub: Yes, the Americans have the skills, experience, and problem-solving ability the industry needs, but thousands of them are forty to sixty years old and *command good salaries.*

Why pay those middle-class salaries when you're Bill Gates or some other big dog who gives big bucks to both parties and can get big favors done for you in return? So tens of thousands of U.S. computer professionals are sitting idle today because Washington has dutifully opened the gates of selective immigration, allowing Microsoft and the rest to flood the job market with H–1B imports who are paid a third to a half less than the going rate, then using the foreigners as a tsunami to knock down the salary scale for the whole industry.

If it was only a matter of finding homegrown talent, Tom Cox would have been snatched up quicker than a frog snatches a fly. He's in the hot high-tech market of Austin, Texas, where computer executives wail ceaselessly that the pool of skilled American labor has gone bone-dry. Yet Cox is toting a résumé heavier than he is, including thirteen years' experience as a top-flight computer programmer and fluency in eight computer languages, from COBOL to Java. But after knocking on corporate door after corporate door, all this forty-five-year-old programmer has gotten is a fat file of rejection letters. As he told the *Austin American-Statesman*, if the companies don't want to hire Americans like him, they should just say so, "But they should quit complaining about a labor shortage, because there are plenty of us out here who can do the job."

People like Cox are being denied the opportunity to "do the job" from coast to coast: Alan Ezer of Queens, New York, has up-to-the-moment skills and ten years of computer engineering experience, but after sending out 150 résumés over the past couple of years, he has been granted only one interview, and no job; Jerry Kowaleski, fifty-eight, is a major semiconductor talent who served thirteen years as a project manager for Texas Instruments, and served more recently as manager of a state-of-the-art semiconductor facility in Thailand, but he has spent the past year trekking from one company to another in a fruitless job search; and Bard-Alan Finlan is a forty-something computer engineer in San Diego with the requisite abilities and credentials to be hired in that high-tech hot spot, yet he, too, has gotten nothing but the runaround, including almost comical treatment at various high-tech "job fairs," which he described to Copley News Service: "You take a pile of 50 resumes and give them to the same companies you gave them to six months earlier at the last fair. They have someone sitting behind a desk smiling at you. They give you a nice plastic pen and a balloon, and you move on to the next desk."

You want class warfare? Cynthia Chavez Wall could have told you about it. But she is dead now, one of the war's casualties. Ms. Wall was a single mother who worked at a textile factory near Hamlet, North Carolina, for thirteen years. She was making $8 an hour until she was abruptly fired one day for not coming to work, having stayed home to care for a daughter, who had come down with pneumonia.

Desperate for a job, she hired on at Imperial Food Products, even though it paid her only $4.95 an hour. She cut up and prepared chicken parts that were sold to fast-food restaurants. She often went home with her hands bleeding from the cuts she inevitably got trying to keep pace with constant demands to speed up the process. She worked up against fryers with oil heated to 400 degrees; no air conditioning, no fans, and only a few small windows. She found it hard, sweaty, dangerous, hellish work. She got thirty minutes for lunch and two fifteen-minute breaks. Complaining about any of this got you nothing but fired, and Ms. Wall had to have a job, so she just had to take it.

Then on the morning of September 3, 1991, women in one area of the plant began to yell, "Fire!" Flames flared and smoke billowed throughout the building, which had no sprinkler system, no evacuation plan, and only one fire extinguisher. As the fire spread quickly, panicked workers raced to the exits, but the people shoved on the closed doors to no avail. All but the very front doors had been padlocked from the outside. Company executives later said they did this to prevent chicken parts from being stolen. Trapped, twenty-five of the ninety employees died in the flames. More than fifty others were burned or injured. Cynthia Chavez Wall's body was found at one of the doors.

"Horrific Accident!," wailed the media. But it was no accident. These people were effectively placed in a death trap by their employers—a death trap that had never once in its eleven-year existence been inspected by safety officials, though it was regularly visited by U.S. Agriculture Department inspectors checking on the quality of the chicken meat. Earlier in the year the North Carolina legislature had rejected proposals to toughen the state's safety regulation, even though the system is so lax that the average North Carolina workplace is inspected once every seventy-five years. Under Reagan and Bush, Washington, too, had cut back on the number of federal inspectors, leaving us even today with fewer than 1,200 to check out 7 million American workplaces.

It gets scant media coverage and practically no political attention, but Cynthia Chavez Wall's deadly experience is no anomaly. More than 10,000 working people a year (about thirty every single day) die on the job. These are not gentle deaths—they are buried alive in trench cave-ins, blown to pieces in refinery explosions, scalded to death by pipe ruptures, asphyxiated by chemical spills, and otherwise unpleasantly terminated. Ours is the deadliest workplace in the industrialized world. Other industrial nations simply do not put up with such corporate laxity, imposing safeguards that would prevent a majority of the death sentences our executives hand down to 10,000 of our people year after year. These deaths are the tip of the iceberg—hundreds of thousands of us also die each year from the long-term effects of our jobs: cancer, brown lung, radiation poisoning, and other diseases.

Two years after Ms. Wall died, two years after the media had scurried away to the next "big story" and the politicians had held their hearings and moved on, a watchdog group called the Government Accountability Project revisited Hamlet and the surrounding area. Imperial is no longer there, but the group found that in other poultry plants, nothing has changed. Assembly-line speedups continue to cause excessive injuries, stifling heat and oppressive working conditions remain, ill and injured employees are forced to stay on the line or be fired, and, yes, doors are still locked from the outside.

Meanwhile, back at the ranch in Washington, Vice President Al Gore and his crack team of "reinventing government" experts have so thoroughly reinvented the Occupational Health and Safety Administration that now (Author's Safety Warning: Do not read any further unless and until you are securely seated) poultry processors are allowed to *inspect themselves*. Of course, they are asked to report any safety violations that they find to OSHA. Imagine how pleased Cynthia Wall would be with this Kafkaesque turn of the screw. The Republicans have an even screwier approach to worker safety—they have begun the elimination of OSHA's enforcement power altogether, intending to make it nothing but an advisory agency that consults with companies that might be interested in instituting voluntary safety programs. Thanks to this bipartisan retreat, OSHA has become more lapdog than watchdog, with the number of workplace inspections down by 43 percent since 1995, the number of serious citations down by 64 percent, and the number of workers covered by OSHA protections down by a third.

When Al Capone ran booze in Chicago during Prohibition, he observed, "When I sell liquor, it's called bootlegging; when my patrons serve it on silver trays on Lake Shore Drive, it's called 'hospitality.'" Likewise, today's CEOs, Wall Street speculators, bond investors, and their assorted agents in government do not care to have their collective actions against us middle-classers and poor folks referred to as "class war." Too, too tacky. So they employ all sorts of comforting euphemisms ("global competitiveness" is the current fave) for rationalizing any and every act that takes from us and gives to them. But it *is* class war, and no matter how they try to pretty it up, it's like putting earrings on a hog—they just can't hide the ugliness.

3

THE MEDIA

LIKE CATS WATCHING THE WRONG MOUSEHOLE

★ LIBERAL MEDIA, MY ASS
★ THE DAY TED KOPPEL LEFT TOWN
★ TALES OF THE TUBE
★ THE REAL MEDIA BIAS
★ SYNERGY
★ THE PEOPLE ARE *REVOLTING*
★ ALTERNATIVES

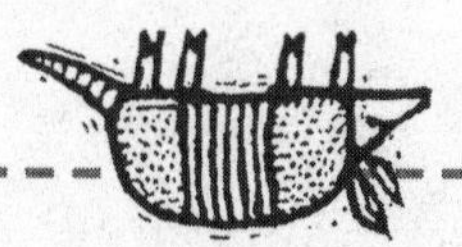

LIBERAL MEDIA, MY ASS

Let us now praise the liberal media: Rush Limbaugh, George Will, William F. Buckley Jr., John McLaughlin, G. Gordon Liddy, Joe Klein, Linda Chavez, Pat Robertson, Bill Kristol, Charles Krauthhammer, Robert Novak, Arianna Huffington, Paul Gigot, Cal Thomas, Michael Barone, Bob Grant, William Safire, Lally Weymouth, Fred Barnes, Ken Hamblin, David Gergen, James Glassman, Pat Buchanan, Bay Buchanan, Kenneth Adelman, Armstrong Williams, John Sununu, Ben Wattenberg, Peggy Noonan, Tony Snow, Lynn Cheney, Richard Perle, Joseph Sobran, Robert Samuelson, Bo Gritz, Roger Ailes, Walter Williams, James J. Kilpatrick, Jeanne Kirkpatrick, Mona Charen, Brit Hume, Ollie North, Mary Matalin, John Leo, Thomas Sowell, Bernard Goldberg, John Stossel, Michael Reagan, P. J. O'Rourke, Jerry Falwell . . .

Liberal media? Give me a break. Sure, there are the delicious bon mots of the untamable Molly Ivins, Bob Herbert's pointed and poignant writing every week in the *New York Times,* Barbara Ehrenreich's insightful essays in *Time,* but contrary to popular misconception, there is a deplorable paucity of actual liberals, much less progressive populists, with access to a national, mass-market megaphone of any kind, either in print or on the air (especially few on the air). So the public gets very little day in and day out to counter the cacophony of voices peppering them from the right.

It is a demonstrably goofy notion that America's media machine is a hotbed of Stalinesque liberalism that spews a steady stream of lefty propaganda from every laptop and lavaliere mike in the land, yet this is a hobgoblin that top Republican politicos have been quick to embrace. Newt Gingrich has been especially cranked up on this theme, whining during the 1996 elections that "the bias in the media is so overwhelming that all the voters hear is the liberal background noise." Bob Dole, too, went around grouching that "We've got to stop the liberal bias in this country. Don't read that stuff. Don't watch television," even though he had been endorsed by nearly two-thirds of the nation's newspapers. Winner of the 1996 "Golden Goofy," though, was Little Stevie Forbes. Despite having inherited his own multibillion-dollar media conglomerate and having his own liberal-bashing column in a mass-market magazine that bears his own name, this '96 GOP presidential wannabe still cried a river during the Republican primary about being mistreated by that old standby, the "liberal media."

Newt tilted even farther over the edge back in the opening days of his speakership, when his star was ascendant and his ego was at full bloat. In March 1995 he urged newspaper owners to clean house, declaring that every newsroom in America was a viper's nest not merely of liberals . . . but of outright *socialists*. Socialists at the *Dallas Morning News*? The *Wall Street Journal*? *USA Today*? The *Peoria Journal Star*? The *Pflugerville Pflag*? The (Insert Your Paper Here)? Name these red devils, Mr. Speaker, name them! Of course, he could not. As Mike Kelley, a reporter with Austin's daily paper (a distant outpost of the far-flung Cox empire), noted: "I don't even like car-pooling."

But the "liberal media" chant makes a good, hot-button campaign schtick, so Newt, Bob, and Little Stevie played it for all it was worth, and then some. Each of them tried to buttress their assertions by thumping various podiums with what they considered a terribly damning survey of Washington reporters, a high percentage of whom identified themselves as—hold on to your hats!—Democrats. None of the GOP complainers whispered a word about the political leanings of the handful of executives who really run America's media show: the general managers, advertising directors, managing editors, publishers, and owners. And while the study on reporters was widely covered, the supposedly liberal media raised no alarms at all about the Tory bias of the bosses.

By trotting out the tried-and-true liberal bugaboo, Newt and the rest divert public attention from those who own and run today's media combines. Ideologically, this establishmentarian clique is inseparable from what it is structurally: Cold, Calculating Corporatist. This group is happy to be either Republican Corporatist or Democratic Corporatist, whatever it takes to perpetuate and extend their domain. At heart, they are devout disciples of the Holy Global Corporate Orthodoxy, distilled to this three-pronged metaphysical essence: Money . . . More . . . Now. This is a secular theology once expressed to me in folksier terms by a West Texas land baron: "I don't want *all* the land, Hightower—just all that's next to mine."

"Corporate media" no longer suffices as an umbrella term for today's multinational, multimedia constructs. The news is no longer in the hands of news companies, much less in the hands of news people. Check the latest conglomerate framework of television, for example: NBC is a wholly owned subsidiary of General Electric, CNN has been swallowed by Time Warner, CBS is under the thumb of Westinghouse, and ABC is now just another product of Disney Inc.

So what—with cable TV there is plenty of room for broadcast diversity, right? Check again, Little Nellie Sunshine. Between them, the Big Four also own the Financial Network, TBS, TNT, Headline News, CNN International, Cinemax, HBO, Comedy Central, E!, Sega Channel, CNBC, Court TV, Bravo, American Movie Classics, MS/NBC, A&E, History Channel, Disney Channel, ESPN, ESPN 2, Lifetime Network, Touchstone Television, Buena Vista Television, The Nashville Network . . . and even the Cartoon Network, among others.

National broadcasting, though, is just one piece of the media industry that GE, Time Warner, Westinghouse, and Disney have in their tentacles. There is a good chance your local TV stations are owned by these conglomerates, too, and they own the licenses of dozens of radio stations. Westinghouse/CBS alone owns eight radio stations in Chicago, seven in New York City, six in L.A., eight in Dallas, and eight in San Francisco—among many others all across the country, giving this one company control of more than a third of the entire U.S. radio audience. Some also own satellite systems, computer networks, and cable TV franchises. They own major newspapers, national magazines, and book publishers (including the Book-of-the-Month Club), making them a power in the print industry as well. Add their extensive holdings in movies, records, theme parks, and sports franchises, and you'll find that a big slice of your information and entertainment dollars is going to just these four.

Liberal media? Bear in mind that broadcasting is not even the main business of the conglomerate holders of our news networks. General Electric, for example, is the country's second largest builder of nuclear power plants and has been a major builder of nuclear bombs. It is also a bank and a major credit card company, the second largest producer of plastic in the United States, one of the biggest weapons makers and arms dealers in the world, a leading stockbroker on Wall Street, a high-ranking Pentagon contractor, a purveyor of insurance and medical services, and a top manufacturer of everything from locomotives to light bulbs. With $70 billion a year in revenues, GE has the *world's fifty-fifth largest economy*—larger than the total economies of such nations as the Philippines, Iran, Ireland, Pakistan, New Zealand, and Egypt.

Whatever else this monolith might be, one thing it damn sure is not is left wing. One chronicler of the company observed it was "so obsessed with conservatism that it was not unlike the John Birch Society." GE, you might remember, is the outfit that sponsored Ronald Reagan as host of the *General Electric Theater* television series in the fifties. In addition, Reagan was directly on GE's payroll as its PR spokesman for eight years, 1954–1962. His "job" was to make radio broadcasts and travel the rubber-chicken banquet circuit, speechifying against commies, unions, corporate taxes, welfare, social security, and all things liberal. It was this sustaining sponsorship that made Reagan the voice and darling of right-wing Republicans, solidifying a core constituency that propelled him into the governor's office in California and ultimately into the White House.

(Lesson Number 14,367G(3)ii on How Politics Works, Entitled "Sponsorship Has Its Privileges": Once the Gipper was safely ensconced in the Oval Office, his old pals from GE came calling, and Reagan's Federal Communications Commission soon began loosening its regulatory grip on television licensing, including making a specific change that allowed General Electric to buy NBC. You see, GE is a convicted felon. Not once, but many times it has been convicted of such felonies as bribery, defrauding us taxpayers, and committing gross environmental and financial crimes. If you had such a rap sheet, you would not be deemed fit to hold an FCC license. But on December 10, 1985, the commission relaxed its fitness restrictions, ruling that felonies would count against a conglomerate buyer of media properties *only* if the company's top executive had been found guilty of the wrongdoing. The very next day GE waltzed through this tailor-made loophole and bought RCA, then the parent of NBC, for $6.2 billion.)

In addition to owning NBC, General Electric also sponsors the *McLaughlin Report*, the inside-Washington gang bang on liberals, which airs on both NBC and PBS. GE does not sponsor any liberal or populist counterpart to McLaughlin's weekly diatribes, and it does not allow any remotely leftish news programming on its own channels—NBC, CNBC, and MS/NBC.

Nor does GE look kindly on any of its media subsidiaries probing the deeds of the parent. Larry Grossman, president of NBC News for the first few years of GE's proprietorship, has related that he literally felt the heavy hand of conglomerate ownership—GE boss Jack Welch poked a finger in his chest and fairly shouted at him: "You work for GE!" Grossman, who later was fired because he balked at the executive suite's meddling ways, soon learned that working for GE means sticking to the company line. A few for-instances:

- In the 1987 stock market crash CEO Welch told his news division not to use the phrase "Black Monday" for fear that it might dampen the company's stock prices.
- Bossman Welch also insisted that *Today* show weatherman Willard Scott be allowed to plug GE light bulbs on the air.
- A 1987 NBC documentary on nuclear power was so pro-nuke that it could have been produced by the industry's PR office, complete with a glowing intro by NBC news anchor Tom Brokaw. Neither Brokaw nor the documentary mentioned that GE makes a killing in the business of building these power plants—a conflict of interest so glaring that offended NBC journalists took to calling the network the "Nuclear Broadcasting Company."
- On November 30, 1989, NBC's *Today* show aired an investigative report about bad bolts used by GE and other firms in building airplanes and bridges . . . except that viewers never saw or heard any reference to GE, since all mentions of NBC's parent were surgically removed by network officials.
- For three nights running in June 1990 *NBC Nightly News* broadcast a gushing series about a new machine to detect breast cancer. The network spent fourteen minutes of airtime on it—the equivalent of *War and Peace* in TV time, where major news stories are lucky to get two minutes. In none of those fourteen minutes was it mentioned that GE makes these particular machines.

- In March 1994 NBC's European cable network canceled a human rights series called *Rights & Wrongs* after it injudiciously produced a show featuring poor working conditions in factories in Mexico—factories owned by GE.

Liberal media? Tune in *Crossfire,* CNN's nightly show that celebrates itself as one television venue where left and right can bare their teeth and go at each other with no holds barred—a kind of verbal mud wrestling for news junkies.

This program was built on a cast of heavies from the right: Pat Buchanan, Robert Novak, Fred Barnes, John Sununu. And on the left? Michael Kinsley. Good grief. Michael Kinsley? It was like sending Tweety Bird into a cockfight. Nice guy. Bright guy. But at best a squishy corporate centrist who makes Bill Clinton look like a standup guy for the working class. Kinsley's idea of class is the class of '72, the year he graduated from Harvard. Isn't Kinsley the writer who opined in *Time* magazine back in '89 that there should be a movement to draft Margaret Thatcher as America's President? He is.

Mercifully, Kinsley resigned the "left's" chair in 1996 to accept a sinecure at Microsoft, putting his faint "liberalism" in the employ of one of America's leading monopolists. At last, though, this meant there was a chance to put someone in that seat who has real teeth to bare. *Crossfire*'s producers led progressive activists to believe that they were serious about wanting a contender, a true progressive scrapper to join the show, so names were submitted. A number of us pushed hard for Jeff Cohen, the founder and director of the highly regarded group FAIR—Fairness and Accuracy In Reporting. Jeff is smart, scrappy, knowledgeable, experienced, articulate, and an unabashed progressive, plus he met a key criterion for the job: He was willing to do it, willing to spend a significant portion of his life in televised shouting matches with the likes of Novak and Sununu. Such is the price of show biz.

The show's producers ran Jeff through the hoops, gave him a couple of tryouts, made positive noises about his professionalism, insisted that he was a finalist, and generally built up the boy's hope. Then they discarded him like a used hankie. It seems that all they really wanted, from the get-go, was someone to defend Bill Clinton.

That's it. Bill Clinton is as far left as the corporatized media is willing to allow the televised right-left dialogue to go, even on *Crossfire,* considered the loosest political format on TV. No Jeff Cohen, who would push the show's right-wing gang to the wall on the job-destroying impacts of NAFTA, on welfare for the privileged, on big-money corruption of the two-party system, on the crushing of America's middle class, on the growing health-care gap between those at the top and America's workaday majority, on . . . well, on the broad range of kitchen-table issues affecting the people—issues that no Clinton defender can hammer, since the President either has gone along with Gingrich Republicans on these issues, or has actually led the fight.

One other issue, too, that undoubtedly dashed any hopes Cohen might have had: the conglomerization of America's media. You can just hear the CNN/*Crossfire* muckety-mucks saying, "Holy Ed Murrow, this guy might take a poke at *us.* Freedom of speech can go too far. Cohen can't be trusted. Next!"

Instead of Cohen (or Christopher Hitchens, Barbara Ehrenreich, or any of several other tried-and-true progressive candidates who would have brought fire, fun, and political integrity to the show), *Crossfire* meekly went with corporate convention, choosing Geraldine Ferraro, another cautious, big-business Democrat. Whatever small spark she might have added was thoroughly dampened by the fact that she was planning from the start to run in the '98 senatorial race for Republican Senator Al D'Amato's seat in New York, meaning she was not about to say anything or take any positions on television that might be offensive to her corporate backers. *Crossfire's* "debate" was perfectly safe in her hands.

OK, her name has some marquee value that Cohen's doesn't, but then *Crossfire* announced it would name a second host to bolster the left's ranks on the show. And the winner was: Bill Press. Hardly a household name, even in his own household, but another safe choice for the powers-that-be. Press was California Democratic Party chairman at the time, totally dedicated to the big-money corporate politics that dominates the party, and, as he wrote to members of the Democratic Finance Council when he was chosen, "thrilled to have the opportunity to defend Bill Clinton and his agenda with my cohost Geraldine Ferraro every night on *Crossfire.*"

In case you are not quite clear on what that agenda might be,

Press explained to the *Los Angeles Times* that Clinton represents a new Democratic effort to reposition the party away from a defender of the have-nots and into a party for the haves: "We have to reshape our agenda and stress the issues that appeal to the haves, like welfare reform and maybe some marginal kind of health reform, but no big, global thing," said *Crossfire*'s latest liberal hope.

On February 19, 1995, Ruth Johnson of Caldwell, Texas, sent this letter to the editor of the *Austin American-Statesman.*

LIBERAL MEDIA. YOU'RE KIDDIN'.

This reminds me of the boxer who took on the champ, because of much insistence from his trainer. After the first round, he came back to his corner with a busted lip and his trainer patted him on the back and said, "You're doing great," then shoved him back in when the bell sounded. Following the second round, he staggered back to his corner with a black eye and a busted cheek and his trainer said, "You're doing great, he hasn't laid a hand on you." With this, the boxer replied, "Well, you'd better keep an eye on the referee, 'cause somebody is beatin' the hell out of me."

THE DAY TED KOPPEL LEFT TOWN

In 1979 I stepped down as editor of the *Texas Observer* to enter politics, thereby making the only downward career move one can make from journalism.

Still, it was worth doing. As I explained in my farewell column, "Sometimes, writing about the bastards is not enough," so I set out across the expanse of Texas to run against them instead, challenging for one of the establishment's "proprietary" offices—the oddly named Texas Railroad Commission, a position that had long been in the pocket of the utility, trucking, and oil industries it regulates.

The state's media dismissed my chances, mainly because I was running against a well-funded and skilled incumbent whose political temperament was summed up by his nickname: "Snake." So I was not altogether surprised to get a chilly reception late one afternoon when I met with the editor of the *Bonham Daily Favorite*. "I've been following you," he told me, squinting his eyes much like a junior-high principal might when addressing an unruly student, "and I can tell you that your attacks on the utilities and oil companies won't get you anywhere here. The people of Fannin County are conservative, and you're just too radical for them."

Even though I felt in my gut he was full of it, I was a bit downcast as I left his office and headed to my next meeting, at the home of a leader in Bonham's African-American community. He had assembled half a dozen other community leaders, and I told them about my lecture from the editor, asking if I was wasting their time and mine in Fannin County. "Oh, I know that man," my host said. "Every morning I get up and pour myself a bowl of Post Toasties and read that man's paper. That way I go to work with nothing on my stomach and nothing on my mind."

We proceeded to organize in Fannin County—and carried it with 60 percent of the vote.

Like the editor in Bonham, the national "insider media" in Washington and New York sit aloof from the experiences of ordinary citizens, largely condescending about the political sensibilities of everyday folks. Those who decide what to report ("All the News

That's Fit to Print," as the *New York Times* so modestly asserts on its masthead) suffer from what we progressive Texans used to call "Morning Newsitis." In my days as a practicing politician, the *Dallas Morning News* was less a newspaper than the official mouthpiece of the Dallas establishment. Its big business and right-wing slant made it journalistically gimpy, incapable of even looking into communities different from its own. As a result, the *News* missed or ignored so much of what was going on in the city that we dubbed it: "The *Dallas Morning News:* If It Happened in Dallas, It's News to Us."

Ted Koppel gave the nation a stupendous display of Morning Newsitis on August 13, 1996, the day he left town. The town was San Diego, the day was day two of the Republican National Convention, and Koppel, the super-sparkly of ABC's *Nightline* show, was piqued because, he exclaimed, no news was happening. "This convention is more of an infomercial than a news event," proclaimed the Emperor of Late-Night News, and off he swept, home to Washington, his battalion of producers, camera people, sound technicians, makeup artists, publicists, and perfumers scurrying alongside him.

Granted, the quadrennial gatherings of both the Republican and the Democratic parties have become minutely scripted, self-congratulatory, made-for-TV extravaganzas, with all the substance, importance, sincerity, and drama of the Miss America contest. But these national conventions are showcases not merely for the pols, but for the pundits as well—not unlike the symbiotic relationship of cows and cowbirds. So the networks assemble all their stars, keeping them prepped and powdered, ensconcing them high above the convention floor in multimillion-dollar sky box studios. The stars then use the conventioneers and the speaker's podium as background noise and visuals for their own pontificating.

Koppel was not the only media show horse embarrassed to be a part of the scene. Tom Brokaw of *NBC Nightly News* penned a cathartic mea culpa for *Newsweek* after the Republican convention, confessing that "In San Diego, I laughed aloud when the only breaking news we were able to report one night was that New York Gov. George Pataki would deliver the nominating speech for Jack Kemp. That received serious, extended analysis when in fact it would have no impact whatsoever—on anything. It was just another example of the self-importance of journalists and politicians alike—and we wonder why so many people hold us both in such low esteem."

Well, yes, but hey, Ted and Tom—call me old-fashioned, but whatever happened to the journalistic principle that news is something one has to go find, rather than sitting around moaning that it's not being handed to you? Yes, the proceedings inside the convention halls were mostly nonnewsworthy nonevents. So: Get the Hell Out of the Hall!

The media are like cats watching the wrong mousehole. Had Tom and Ted really wanted a story in San Diego, they could have taken their mikes and cameras just a short stroll down the street to a hotel marina where the *Quiet Heart* was docked. Aboard this seventy-five-foot yacht they would have found Representative Tom DeLay, the Republican whip, known in Washington as the "Hammer" for his less-then-subtle technique of raising money from corporate lobbyists in need of legislative favors.

Throughout 1995 and well into 1996 the Washington media were riveted on the ideology and personality of Newt Gingrich and his "revolution," as embodied in the Loudspeaker's widely ballyhooed "Contract with America." Stripped of its rhetorical bells and whistles, though, it really was a Contract with *Corporate* America—a veritable wet dream of such long-held fantasies as gutting environmental protections, eliminating worker-safety rules, widening tax loopholes, preventing consumer lawsuits, and granting more corporate subsidies. Newt's "revolution" was not motivated by ideology but by raw politics, fueled by the old quid pro quo of corporate-cash-in and corporate-benefits-out. The man in charge of collecting the cash was the third-ranking Republican in the House leadership, Tom DeLay.

A former pest exterminator, this Houston-area congressman proved good at ledgers, literally keeping book on the 400 largest corporate PACs, listing the amounts and percentages of their campaign contributions that went to Republicans and Democrats, and labeling the PACs as "friendly" or "unfriendly," depending on how heavily they gave to the GOP. If your corporation or industry group wanted a piece of the revolution, your lobbyist had to "go see Tom," who would whip out his book and examine your campaign-giving history right on the spot. If it was not up to snuff, you had to make it right—that is, start shoveling checks into the designated Republican pockets until Tom said, "OK." While Newt pledged publicly to end business as usual in Washington, DeLay's job was to implement a behind-the-scenes plan of business *more than* usual.

At the San Diego convention, Koppel, Brokaw, and other media heavies could have given viewers some up-close-and-personal shots of the key players in this slimy system. They could have stationed themselves at the private landing strips to catch top congressional Republicans deplaning from corporate jets. They could have driven up Interstate 5 to the Torrey Pines Golf Club, where AT&T hosted a private outing for Senator Orrin Hatch, House budget committee chairman John Boehner, and a host of other GOP duffers, including our man DeLay.

Then they could have traipsed after the Hammer as he rode in his limo back to the marina and bounded aboard the *Quiet Heart*. Reporter Damon Chappie of *Roll Call*, a probing little paper that covers Capitol Hill, did track DeLay to the yacht and found the story that Koppel and Brokaw ignored. There was the Republican whip, decked out in white polo shirt and aquamarine shorts, making merry with a boatload of "friendlies," sipping gin-and-tonics, enjoying a lavish spread of hors d'oeuvres, and "talkin' revolution." This would have made an *excellent* television moment, don't you think, a revealing vignette of money and power rubbing up against each other—in their shorts, no less!

If Tom and Ted were too busy to catch the cocktail event on *Quiet Heart*, they could have caught DeLay on board the *High Spirits* for a breakfast fund-raiser he hosted for still more favor-seeking lobbyists, or on the *Renown*, yet another yacht he used during convention week for even more corporate fund-raising. The breakfast event on *High Spirits* would have been perfect for TV, cohosted as it was by Nick Brookes, the CEO of British tobacco peddler Brown & Williamson. Given that Republican nominee Bob Dole proved such a standup guy for the nicotine industry throughout his presidential campaign, and that late in the campaign he also flogged Bill Clinton mercilessly (and deservedly) for being in bed with foreign-money interests, the British tobacco connection to Dole's San Diego convention would have been both an entertaining and an informative bit of journalism.

OK, investigative journalism can be tiring, since it involves, well, investigating, but it's not like DeLay was laying low, trying to hide from the probing eyes of media sharpies. To the contrary, as he dashed from yacht to yacht, he giddily exulted to *Roll Call*'s reporter, "We're raising money like crazy!" On Tuesday night he even staged a garish yacht-fest, docking all three luxury boats at the marina, where he loaded them with lobbyists and members of Congress. Then he set sail across the harbor to Coronado Island for an evening of schmoozing and boozing at a posh restaurant.

Wouldn't it have been great if Tom and Ted had loaded their camera crews in a couple of speedboats and roared off in hot pursuit? The two news stars could have donned yellow slickers and stood in the bows of their boats, courageously thrusting their microphones at DeLay and his gang of special-interest pirates. Now *that's* TV! And news.

But Ted was packing, and Tom was drafting his *Newsweek* piece, so all the public saw were more talking heads inside the convention hall.

TALES OF THE TUBE

Ralph Kramden: Look, Alice, it ain't just the money. There's another reason I don't want no television in this house. It changes people. They stop usin' their brains. They just look. When people get television, they stop readin' books.

Alice Kramden: Well, it can't change us. We don't even have a book!

Ralph: All right. All right. I'll get you a book!

The Honeymooners
From *Primetime Proverbs*

In politics, television has more than changed us, it has become our reality, more real in fact than life itself.

This was brought home to me in 1989 when I began my reelection campaign for state agriculture commissioner and was in the West Texas hamlet of Paducah for a little fund-raiser thrown by some area farmers, lawyers, merchants, and such. Thirty or so folks gathered in the home of one of the farmers, chatting amiably, enjoying a couple of beers and grazing on the usual Texas campaign spread of chips and hot sauce, queso and tamales. Then, around 6-ish, the hostess asked us all to convene in the living room, presumably to hear me offer a few thousand well-chosen words and show myself witty, wonderful, and wise, well worthy of their $25-a-head contribution to my political prosperity.

But no! As I shuffled into the living room behind the group, pondering just the right snappy line to open my comments, I hear the hostess saying, "Rather than hear Jim speak, we were able to tape the show that *60 Minutes* ran on him recently, and we're going to play it for you." So there I stood, the actual candidate, live and in person, cooling my heels for fifteen minutes, while my supporters turned their backs on me and gaped at my electronic image, far more impressed by it than by me in the flesh.

Actually, I learned about "television reality" a decade earlier when I first ran for office and found to my dismay that even ardent supporters did not consider me a real candidate unless and until they saw me on the tube. This credibility threshold is even more rigidly demanded by the media gatekeepers themselves—to give you coverage as a "real" candidate, television assignment editors and the big-city press insist on knowing how much coverage you have received, a Catch-22 that helps explain why so many political candidates are loopier than a dry-land farmer in a four-year drought.

To have a prayer in politics today, we underfinanced populists (probably a redundancy there) must become adept at drawing free TV coverage, and/or hoard enough of our meager purse to buy one or maybe two TV spots toward the end of the election. These paid ads obviously are not enough to shift any undecided voters, but are bought simply to buck up the spirits of those already for you—"Hey, Nadine, come on in here right now! Ol' Hightower's got himself a tee-vee ad, look at this!" We would even mail notices to our supporters, alerting them to the schedule of our lonely ad: "Be sure to catch our sixty-second spot to run at 5:17 A.M. next Thursday on Channel 11's *Early Bird Farm Report,* and call all your friends, too, and . . . "

I have to tip my hat to Paul Wellstone, though. He gets the Emmy for Best Use of an Ad Campaign That Never Aired Anywhere. Wellstone is the out-and-out liberal college professor and all-around fine human being from Minnesota who ran a hardscrabble, seat-of-the-pants campaign in 1990 and won a U.S. Senate seat. Paul's "media strategy" was to do a Michael Moore–style TV ad in which he and his camera crew went in search of his opponent, a multimillionaire Republican incumbent who had an ethical rip or two in his senatorial toga and refused to debate this hard-charging challenger. The ad was a winner—pointed, humorous, and memorable.

Problem was, Paul had no money to pay the TV stations to run the damn thing. So what he does is call a press conference and show it. The ad is so funny and telling that it gets airtime all over Minnesota, nailing his opponent and thrilling his supporters, who, at last, see him on television.

In 1976 I used the media's own tube fixation to my political advantage. I was national coordinator for the presidential campaign of the populist U.S. senator Fred Harris (I like to brag that I made Fred what he is today—a professor at the University of New

Mexico). Nonetheless, we had our moments in that campaign, and one of them was in the Iowa caucuses.

The '76 Democratic primaries were crowded with challengers, including Mo Udall, Scoop Jackson, Birch Bayh, Sargent Shriver, Jerry Brown, and a couple of others, but this was Jimmy Carter's year, and he was way ahead of us in Iowa. Our task, however, was not to beat Carter, but only to "beat media expectations," to run better than the pundits thought we would. As luck would have it, the media, with their usual dismissal of underfunded populists (there's that redundancy again) had ranked Fred beneath "other" in their preelection pontifications.

This played perfectly into my Machiavellian "strategy" of placing third or fourth in this first-in-the-nation vote, thereby getting a media boost from stories that would say, ". . . and Fred Harris did surprisingly well in Iowa, running ahead of expectations . . ." This could help our fund-raising and energize our supporters as we headed to the New Hampshire primary.

Fred's populist message, humor, and dynamite speaking style drew a good response in Iowa, and despite having practically zero money for TV ads, we felt he *would* run third or fourth in the Monday night caucuses. There was a problem though: The caucuses were a new creature in presidential politics back then, and there was no computerized statewide system to gather the results from hundreds of these citizen gatherings and report them quickly to the national press corps, which operates on East Coast time and had a 9 o'clock deadline to get anything in the *New York Times*, the *Washington Post*, and the rest. So we faced the age-old question: If you do "surprisingly well" in the Iowa caucuses and the media aren't there to report it, does it make a noise? Not in today's media-centric politics.

So, with our field coordinator Barbara Shailor and our media coordinator Frank Greer (it was a populist campaign, so everyone was called a coordinator, a snappy title that helped us forget we weren't being paid real money), we came up with our own reporting system. From hundreds of precincts across Iowa, our precinct coordinators were instructed by Barbara to jump on the phone the instant they had results and relay them to our storefront headquarters in Des Moines. Frank had quietly notified key media players that ours would be the only place in all of Iowa where they could get vote tabulations in time for their deadlines, and sure enough they took the bait (like fish snapping at nightcrawlers, sports reporters and political reporters are two species that find any promise of numbers an irresistible lure).

But how to make our numbers credible? Simple: television.

If I, as the campaign chief of one of the contestants, had brazenly tried to hand out results phoned in from our own partisans, well . . . even the national press corps, though used to spoon-feeding, might have balked. Luckily, though, our storefront office (with a long front room and a small back office) was just right for our plan. We put a curtain over the door between the two rooms, put Barbara in the back with a bank of phones and a clerk at a computer. In the outer room we put a TV screen on a fold-up table and stationed Frank beside it. On election night, as calls came in from the caucus meetings to Barbara's phone bank, she would pass them to the clerk, who typed them into the computer, which was connected to nothing but the TV screen some twenty feet outside the curtain. The numbers spewed onto the screen, and Frank was there, ever so helpfully, to interpret them for the assembled media.

What a sight! There stood big-name reporters from all the brand-name news outlets, excitedly eyeing the tabulations as they flashed on the screen and compulsively jotting notes in their pads, just as though these were "official" results.

We were able to put up tabulations from maybe 20 percent of the caucus meetings before the media's deadline, and while we did not in any way mess with the actual results from the caucuses, they *were* our numbers, and these were the ones reported. The next morning, while other campaigns were still waiting for the state party to provide official results, news stories across the country read, ". . . and Fred Harris ran surprisingly well in Iowa . . ." What a difference a television screen makes.

Lest you think campaign coverage since 1976 has become one whit more sophisticated or dedicated to digging out anything remotely connected to real news, I invite you to peek inside 1996's first presidential debate in Hartford, Connecticut, between Bill Clinton and Bob Dole. This peek is courtesy of Dan Perkins, the marvelous editorial cartoonist who works under the nom de plume

Tom Tomorrow, and who had the surreal experience of watching the media watch the debate.

Most of the journalists who covered this widely hyped event were not actually allowed inside the Bushnell Theater, where the two debaters faced off, but instead were corralled many blocks away in a specially arranged "media center," viewing the thing on TV. Amazing. Hordes of political reporters traveled from as far as 3,000 miles away at great expense to watch the debate *on television,* which they could have done at a local bar back home.

Not that they were without libations in Hartford. Perkins reports that the media center was catered courtesy of Philip Morris, which also doled out to the media-ites all kinds of trinkets from the company's subsidiaries, including toy whistles shaped like the Oscar Mayer weinermobile, "limited edition" boxes of Kraft Macaroni & Cheese containing noodles shaped like Democratic donkeys and GOP elephants, and even handy reporters' notebooks festooned with the P-M logo. Perhaps sheer gratitude for the grub and the toys accounts for the failure of any of the major-media attendees to mention in their reports the irony that Philip Morris, a major corrupter of the political and legislative processes with its millions of dollars in self-serving contributions to both political parties, was this night's official keeper of the media.

The "news" from the Hartford debate also was catered. Once Clinton and Dole had done their bit and the TVs were turned off at the "Philip Morris media center," spinmeisters from the two parties (no Perot or other third-party people were allowed within shouting distance of the place) suddenly appeared in the room, strategically placed so these supposedly self-respecting journalists could mob them and be told what to think about what they had just seen on the telly.

There were George Stephanopolous, Haley Barbour, Donna Shalala, and others—take your pick. The place looked like some bizarre political bazaar, as each of the partisan spinners had an aide standing beside him or her holding aloft a laminated red sign emblazoned with the spinner's name. Each was quickly surrounded by a swarming mass of card-carrying, professional reporters intently shouting such penetrating questions as: "Who won?"

This sad demonstration of big-league political journalism did not let up even after the spinners had spun their webs and departed the hall. Here is Dan Perkins's personal narrative of the final scene: "As the crush of people around me begins to lighten, I see that there are now reporters interviewing reporters. Someone from a local radio station interviews me. I realize I have fallen into the belly of the beast, this strange, self-contained world of political reporters who travel around in packs, are spoon-fed press releases, spend a frenzied hour gathering sound bites like children on an Easter egg hunt, and call it all news."

THE REAL MEDIA BIAS

I won Marty Nolan's respect in 1976, when he was covering the presidential campaign for the *Boston Globe*. It was not my strategic brilliance or my witty political repartee that impressed him—it was the place I took him to lunch.

Actually, he was taking me, putting our tab on the *Globe*'s account, but I got to choose the restaurant—a nice little gesture he made to all the campaign managers that year. The other managers clearly were sharper than I, though—Marty mentioned as we sat down that the others had all chosen the most expensive restaurant they could think of. Me, I met Marty at the Post Pub, a working class bar and grill where we each had a Ballantine Ale and a baconburger, as I recall, and, of course, fries. Good grease, as we say in Texas. Marty was ecstatic—apparently he was sick of escargot in burgundy sauce, and this pub was his kind of joint.

The surprise is not that I was there, but that a media big-leaguer was. Alas, Marty Nolan and a few others like him are the exceptions, throwbacks to a time, not all that long ago, when nearly all the news was the product of rumpled working stiffs who pounded a beat, drew a modest paycheck, ate lunch at pubs, drank after hours at places with names like Shorty's, and identified with other working folks.

Today, though, most of our news is just another consumer product of the conglomerates that control the media machinery, delivered to us by an upscale caste of finely tailored, college-educated journalists and managers who drink at exclusive clubs like the Headliners or the Uptown, never darkening the door of shot-and-beer joints like Shorty's.

The gatekeepers of our news, like the politicians who claim to represent us, have moved out of our neighborhoods and now are among the wealthiest 10 percent of Americans, identifying a whole lot more with the hoity-toity than hoi polloi. Personally, they care less about the affordable-housing crisis than they do about making sure the deductibility on vacation homes is protected; they see NAFTA in terms of a dinner party discussion about free-trade ideology, not as an imminent threat to their jobs and wages, as it is for

America's middle class; to them, universal health care is a political issue, not something you actually need right now, or fear you will need next year. Their world—and consequently their worldview—is widely separate from the workaday majority's. As a result, the masses get a daily news feed that is neither of, by, nor really for the masses—those 75–80 percent of Americans who do not have a college degree, who are paid less than $50,000 a year, who are not enjoying the present "prosperity" that both the politicians and the media keep talking about, who believe that Washington serves corporations and the wealthy and does not give a fat fig for folks like them, who are looking for a third (or fourth or fifth) political party, and who either are not voting or are voting against the system.

The true media bias in our country is not to the left or to the right, but to the top. It is a bias that filters our news through the lens of the privileged and leaves most families feeling as shut out of the public debate by the elite media as they are by the elite politicians. It is also a bias that often reveals the news presenters to be embarrassingly out of touch with terra firma.

One such moment came on September 17, 1992, when Diane Sawyer used her *Prime Time Live* pulpit to scold a single mother who was trying to keep her family together by working two piss-poor, part-time jobs while also taking a monthly welfare check of about $600. Since this poor mom was "employed," Ms. Sawyer preached, she was not entitled to welfare and, indeed, was "gouging the taxpayer." On a roll, the TV Lady told the Welfare Lady: "You know, people say you should not have children if you can't support them." Now Diane Sawyer, whose ABC contract puts something like $7 million a year into her finely feathered nest, is paid more each and every day than that mother makes in a whole year, and if it is taxpayer gouging that concerns her, how about a report on the assorted loopholes that the privileged cut into the tax code so pampered ones like Ms. Sawyer can have their accountants shelter income for them, shifting her tax burden onto the backs of the middle class?

If she wants to interview a real welfare moocher, her *Prime Time Live* partner Sam Donaldson would be fun! He is knocking down about a million-and-a-half a year with Disney/ABC plus getting paid $25,000 to $30,000 a pop to give speeches to insurance executives and other corporate groups. He has become such a wealthy newsman that he can afford to moonlight as a gentleman rancher near Ruidosa, New Mexico, where he raises sheep. Actually, he does not raise the sheep, he has a ranch manager and a crew of hired hands to do that. And while Donaldson does own the Chavez Canyon Ranch, a big part of his spread actually is state land, some 2,300 acres of which he leases for the subsidized rate of about a buck an acre, roughly eight to twelve times cheaper than the going rate for leasing private ranchland in the area.

In addition to enjoying cut-rate grazing, courtesy of New Mexico taxpayers, Sam the Sheep Man also has rummaged in the pockets of us federal taxpayers, taking about $100,000 in subsidies for his ranch's mohair production. Even stranger, Uncle Sam protects Sam's sheep from coyotes and other critters. You read it right: *We* protect this millionaire's sheep. He signed up for a little-known freebie that sends government-paid hunters and trappers out to his Chavez Canyon Ranch to kill any predators they can find, thus letting Sam and his sheep sleep better at night. According to Citizens Against Government Waste, a watchdog group, the U.S. Animal Damage Control Program has dispatched agents out to Donaldson's spread 412 times in the past five years, killing seventy-four coyotes and three bobcats at a cost of tens of thousands of dollars.

I don't know about you, but I cannot find words to express to Congress how grateful I am that it is cutting back on Medicare, Head Start, job training, and other federal spending so our tax dollars can now go to such essential national needs as killing coyotes on Sam Donaldson's ranch. This is an absurdity that caused longtime New Mexico environmentalist Pat Wolff to cry out in exasperation: "It's outrageous to expect taxpayers to pay government agents to kill wildlife on public land leased by a millionaire TV star playing rancher."

The arch-browed TV personality, always sensitive to the slightest perception of impropriety in the actions of others, sniffs that his subsidies are simply a matter of entitlement. He told the *Albuquerque Journal* that he has a right to all the taxpayer dole he can grab: "The government helps ranchers and farmers and businesses of all kinds. If it is in existence and I am eligible to use it, I'll use it."

SYNERGY

My heart was racing faster than a new sewing machine that August morning, 1967, when I reported for my first day of work as a legislative aide to U.S. Senator Ralph Yarborough of Texas. Barely twenty-four, bright-eyed and bushy-tailed, I strode down the magnificent marble halls of senatorial power and prestige at 8 A.M., swung into Yarborough's suite of offices, walked back to my assigned desk crammed in a corner of the legislative office, took a deep breath . . . and marveled that I was actually *here,* a country boy only two years out of North Texas State, now in a position to influence a United States senator, to write laws, to—oh golly—*make history*!

Part of the duty of each legislative aide was to respond to constituent mail on issues in our particular bailiwick. On my first morning of my first day I opened my folder and took out my first letter, ready to engage in the weighty matters of state with this honorable citizen who had taken time to write to his senator. It was addressed: "Dear Peckerwood."

So much for my illusions about being involved in statesmanship.

I recalled that "Dear Peckerwood" salutation on August 4, 1995, the day the U.S. Senate passed the Telecommunications Deregulation Act. This monster of a bill promised to unbridle America's media giants so they could merge and become even gianter giants, thereby allowing them to increase telecommunications competition, deliver more consumer choices in both print and broadcast journalism, lower prices, create millions of new jobs, put a chicken in every pot, stop dental plaque in its tracks, and build a bridge to the twenty-first century. Wait, that last one was Bill Clinton's reelection promise, or was it? Never mind, none of it was true anyway.

We quickly got the gianter giants—cable powers merging with broadcast networks, networks merging with show-biz empires, telephone companies merging with one another and then remerging with the already-merged cable powers, show-biz empires, and broadcast networks. The diagram of the electrical circuitry inside the space shuttle cannot be as complex as the organizational charts of these hydra-headed media giants, which are still absorbing one another and becoming more gigantic by the moment.

This only makes me want to shout: Hey, peckerwoods, where are all those benefits we were supposed to get from this? Since the bill passed, our phone and cable rates have gone up; instead of new jobs, each merger has produced massive job cuts (in 1996, after Ted Turner's cable empire merged into Time Warner, he fired 1,000 employees, including his own son, for chrissake—he took Ted IV to dinner right after the merger and told him, in these exact words: "You're toast"); since companies that used to compete have merged, there's less competition, not more; and not even the industry still claims that merger mania has produced more news choices for us, much less better journalism.

Far from presenting a new diversity of viewpoints on the airwaves or digging into meaty stories that the premerger media missed, the new amalgamations are into what they call "synergy." This is a 50-cent word that the dictionary defines as: "1. combined action. 2. the cooperative action of two or more muscles or the like." You would define it as: "1. shameless self-promotion. 2. stupid." In practice, what synergy means is that one division of the conglomerate scratches the back of another division, and vice versa. For example, only eleven days after Disney took over ABC, the network's *Roseanne* show just happened to feature a two-part episode of the show's characters going to Disney World! In case *Roseanne* viewers missed the subtlety, Disney World also advertised on the show. Follow the bouncing ball: A Disney ad sponsored a show about a Disney theme park, all of it airing on a Disney network. Disney, Disney, Disney. Synergy.

Our daily news gets caught up in this synergistic fun, too. In 1996, for example, Disney World gave itself a week-long party in Orlando to celebrate its twenty-fifth anniversary. Broadcasting *live* that week from the Florida amusement park was *Good Morning America*. Not coincidentally, *GMA* is the morning news report of ABC, which itself is owned by Disney. To make it seem more "newsy," the *GMA* crew broadcast a special feature about a new ride at Disney World, dutifully reported by ABC's science editor.

Synergy works in reverse, too, keeping items out of the news that owners do not want covered. Freelance investigative writer Mark Dowie experienced this recently when he got "moused" on a story he was preparing for *Los Angeles* magazine. According to a chronicle of his saga in the *Nation,* Dowie was assigned to do a lengthy article on the powerfully conservative Chandler family, which owns the *Los Angeles Times, Newsday,* the *Baltimore Sun,* the *Hartford Courant,* and many other publications, as well as lots and lots of land in L.A. and throughout Southern California.

Family patriarch Otis Chandler was not charmed by the attention, but *Los Angeles* was an independent journalistic product, so reporter Dowie kept doing what real reporters do—digging into the story. While he was digging, though, *Los Angeles* was sold. To Disney. Next thing Dowie knew, a new editor was at the helm of the magazine, and his piece on Chandler had been axed. Disney's new editor told Dowie that it "just didn't fit."

Didn't fit what? Dowie did not get an honest answer until a year later, when he read in another magazine that Otis Chandler was bragging he had killed the story with a single phone call. Chandler said that just before Disney bought *Los Angeles* (the magazine, not the city . . . not yet), he called the top dog at Disney, Michael Eisner, and asked him: "Do you realize you're going to do another—*another*—piece on the Chandlers?" According to Otis, Eisner replied, "Don't see why." And that was that.

Spiking specific pieces is not the worst damage done to journalism by these conglomerations. Far more threatening is their corporate consensus—a sort of unspoken gentlemen's agreement—on what merits (and, more significantly, what does *not* merit) mass media attention, giving this handful of for-profit, private entities awesome control over our democratic discourse. "Not to worry," the media ministers assure us, asserting they are citizens, too, and that they take their responsibilities seriously. "Besides," they say, slapping the table as if this shuts the case, "we're competitors; if one of us doesn't cover a story, another of us will jump on it and beat the others' brains out! That's how competition works."

So, where were these "competitors" during the passage of the Telecommunications Deregulation Act itself? This was the most far-reaching piece of legislation of the 1995–1996 Congress, affecting people more profoundly than Bill Clinton's widely covered welfare deform bill, for example. At issue were not merely our monthly telephone bills and our television fare, but the very dissemination of news and information that a free citizenry must have. This bill represented a powerful extension of the old reality that freedom of the press is limited to those who can afford to own a press.

Where were ABC, CBS, NBC, TBS, Fox, or any of the other television "competitors" on the need to lay this big honker of a bill out for "We the People" to examine? Sure, it got some coverage in the print media, but nowhere near commensurate with the magnitude of the changes it posed, and what coverage it did receive was mostly relegated to arcane articles in the business section, or cast as an internal industry flap between the long-distance phone companies and the "Baby Bells," the local phone companies.

Where was the media on the biggest story in modern times *about* the media? Industry apologists say that the story was too complicated, too difficult (especially for television) to make understandable, too boring for people to follow. This from an industry that devoted *hours* on television and *miles* of column inches in newspapers and magazines to explain the DNA matchups between O.J. Simpson's blood-flecked socks and the bloody glove found behind his house, an industry that put all its technical wizardry and all its reportorial skills to the task of making DNA science comprehensible, an industry that put hundreds of reporters and expert analysts on the story and even moved its network anchors out to "Camp O.J. " adjacent to the L.A. County Courthouse so they could all broadcast live from the scene. Why did they not apply a tenth, a thousandth, a ten-thousandth of this media horsepower to a story that seriously matters?

Where was the media? Hiding the story, that's where, because their conglomerate chieftains and chieftains-to-be were poised to make billions from passage of the legislation, and they were not about to risk getting the public involved.

Indeed, the Westinghouse buyout of CBS, which was announced just before the passage of the Telecommunications Act, was *illegal* under the law then in effect. It could only become legal if this tailor-made telecommunications bill passed, so one of the biggest media mergers in history literally was hanging in the balance. Where were CBS's competitors? Why did *no one* report this slime-covered, special-interest, conflict-of-interest piece of stink? It is a gross case of journalistic malpractice. With investment bankers, Wall Street speculators, merger consultants, corporate lawyers, and assorted media executives standing by to cash in on the merger frenzy, this was not an industry interested in trusting democracy by reporting the news.

Neither was it interested in trusting its own reporters. According to *EXTRA!*, the excellent newsletter of Fairness and Accuracy In Reporting, ABC News planned a report in 1995 on the stockholdings of Representative Thomas J. Bliley—stocks that revealed embarrassing conflicts of interest with his position as chair of the House commerce committee. "Whoa there, back off, Bubba," came the alarmed cry from ABC's top echelons. Bliley, you see, was chair of the committee handling the telecommunications bill, and he did not take kindly to journalistic snooping, so he simply rang up ABC's chief lobbyist in Washington. That *would have been* that, except a brave soul at ABC threatened to blow the whistle on the suppression, so a story did air, albeit a thoroughly watered-down version of what the news division had planned.

If one of the media "competitors" had really wanted to scoop the others and present a dynamite story on the special-interest rewrite of telecommunications law, all it had to do was point a camera down practically any hallway in Congress, which swarmed throughout 1995 with lobbyists for Disney, GE, Fox, Cox, Time Warner, TBS, TCI, Westinghouse, and every other media powerhouse, all of them urging lawmakers to ram this stinker through before the citizenry could get a whiff of it. Or they could have caught Michael Eisner, Ted Turner, and other *jefes* of the new conglomerate journalism coming out of the White House, where they had been tending to Bill Clinton, keeping him properly greased and on track. Or they could have run a delicious series focused on the $40 million that the industry gave to lawmakers, Democrats and Republicans alike, to pass the very legislation that its "news" divisions were hiding from us folks. Here was an Emmy Award–winning opportunity for someone, a chance for the media to shine. But all the "competitors" took a pass.

The gentlemen's agreement assured that there were no telecommunications on the Telecommunications Act itself, and we can expect more of the same on other big issues, where the financial interests of the conglomerates supersede the public's interest.

THE PEOPLE ARE *REVOLTING*

February 1996: Another political upstart rallies a ragtag army of followers and strikes the imagination of millions of America's political outcasts. Tapping into our country's deep and spreading pool of populist anger, he could be trouble. Quickly the flustered establishment rushes out its media sharpshooters, who start firing wildly at the upstart: "wacky," "simplistic," "misleading," "anticorporation," "nonsense," "perverse."

Whew! All these shots were taken in a single six-inch *New York Times* editorial following Pat Buchanan's startling second-place finish in the 1996 Iowa caucuses. So many invectives, so little space—to your thesauruses, media writers, reload and fire, for the infidel Buchanan is charging full-tilt at the Holy Walls of the Established Order! Indeed he was, hurling fiery bolts of outrage at NAFTA, the Fed, excessive executive salaries, the New World Order, and other vulnerable buttresses of America's economic hierarchy.

Then, a week later, Buchanan *won* the New Hampshire primary and, oh, Katy Bar the Door, the media went into full, Code Red assault: *Newsweek* ran a cover article with a dark, two-page shot of Buchanan's face glowering at the reader and a bold-type headline warning "Extreme Measures"; the prissy establishmentarian George Will opined on the air and in print that with Buchanan, "Conservatism gets soiled"; even Rush Limbaugh, the 800-pound gorilla of the Republican power elite, rumbled onto the air the day after the New Hampshire victory, wailing at his listeners: "I'll tell you something, you are being manipulated in a way that I find very bothersome. Pat Buchanan is not a conservative. He's a populist."

Well, actually, Buchanan is not a populist—he is a xenophobic millionaire right-winger and Washington insider who, even while bashing Wall Street and the global money lords, had AT&T, GM, Wal-Mart, China Light & Power, Westminister Bank of London, and other offensive stocks in his personal portfolio. But he *had* learned to strum some true populist chords, and this is why he was both scoring big numbers at the polls and scaring the bejeezus out of the Big Boys. Hence their all-out bombardment.

America's corporate establishment and the mass-market media, now one and the same, would not have been the slightest bit perturbed had Buchanan kept his wrath (and that of his followers) focused on "Mexican border-crossers, welfare cheats, uppity women, baby killers, homos, big gubbmint liberals, Godless schoolteachers, dark-skinned people taking white people's jobs" and all the rest of his standard litany of horrors, because all of these are social issues that pit the powerless against the powerless. Buchanan himself had been especially skilled at this bitter art from the early 1970s when he penned poisonous prose for Spiro Agnew and on into the 1990s during his tenure as the right's near-rabid attack dog on CNN's *Crossfire.* But in the '96 campaign, Pat discovered the working class—the deep angst among America's majority about their falling living standards, the global squeeze on wages, and the rising inequality in the land of "justice for all." Suddenly Buchanan was talking poignantly and convincingly about the Great Unspeakable in American politics: *class.* His message was, as Limbaugh detected, recoiling at the thought, populist. Plus it was succeeding, even in the *Republican* primary, surely the least fertile soil there could be. This made those at the top nervous as a pig in a packing plant, and there was only one thing to do: Kill the messenger.

The same thing was done to Ross Perot in '92, and the media's guns would again turn on him in '96. Focusing on the messengers, though, missed the real story. The significance of both the "Perot Phenomenon" and the "Buchanan Phenomenon" was never the two candidates, but the phenomenon, the fact that there is a seething mass of Americans shouting at the establishment's tone-deaf, two-party system, "You are unadulterated horse-hockey!"

Even though Perot and Buchanan are two terribly flawed political buckeroos—ranging from ludicrous to loopy on some issues, mostly wearing the cloak of populism rather than *being* populists—they drew powerful support, not because their supporters were nuts, duped, or even "supporters," but because both were the handiest two-by-fours that folks had available for whacking Washington and Wall Street upside their thick heads. The media portrayed Buchanan and Perot as "using" working-class resentment, but I can tell you from my political travels and my conversations on talk radio that the exact opposite is true—the people were using the candidates. These two messengers were easy targets for the media, but there will be other messengers tapping ever more effectively into the volatile, powerful, and growing American majority that Washington and Wall Street, the two parties, and the media want to pretend is not there.

From the moment the polls closed on the '96 elections, the media went around plastering little yellow Happy Face stickers on the results, proclaiming confidently that people *really* are satisfied with the way things are, that there *really* is not that much anxiety anymore about corporate downsizing, that big-money corruption of politics *really* did not bother voters all that much, that populist appeals *really* do not work, and that things *really* are perfectly OK. Really.

On *This Week with David Brinkley,* the tightly bow-tied George Will pronounced that the issue of downsizing was so much nonsense, "magnified" by political opportunists operating under the misguided notion that "it is the job of the corporation to be a mini-welfare state." Likewise, *Newsweek*'s head-in-the-sand economist Robert Samuelson intoned that workers' insecurities were "exaggerated" in the election. Typical of the postelection analyses was *USA Today,* which asserted matter-of-factly that Clinton had won because "This year the economy is generally regarded as good."

Oh? By whom? By Al Neuharth and the other $1,000 suits running *USA Today,* for sure, but who else? Well, the paper reports, "Most voters, 53 percent in Election Day exit polls, think the country is moving in the right direction. Clinton got credit for that." Even accepting that the 53 percent, feel-good number is accurate, it still suggests that the other half of exiting voters do *not* think America is heading in the right direction. There was no mention of them. Nor did *USA Today*'s analyst mention that more than half (100 million-plus) of eligible Americans took a whiff of Bill and Bob on Election Day and said, "I don't think so," skipping the voting booth altogether, making 1996 the lowest turnout since 1924 and the second lowest since 1824.

This nonvoting majority is a long way from thinking everything is hunky-dory, as polls have made clear and as any visit to any working-class hangout in the land will make even clearer. You can add to them the 9 million voters who cast their ballots for Perot, Ralph Nader, Harry Brown, John Hagelin, Howard Phillips, and other third-party candidates. Also add in the 15 million Clinton voters who told exit pollsters that they were not voting *for* him, but against the others, only going for Bill because he was the "best of a bad lot."

Connect the dots of these 125 million voters (nearly two-thirds of those eligible) and you will not find a picture of a smiling, satisfied electorate, but a brooding civic rebellion. The breadth and depth of this disaffection from the two-party establishment is the most profound political story of the decade, but the media overseers are either too far above it to notice, or too incorporated within the establishment to be willing to dive into the story (with the admirable exception of the *Philadelphia Inquirer*'s ten-part series by Donald Barlett and James Steele). Confronted with the fact of a historically low turnout in '96, the response of the media establishment was typified by one N. Don Wycliff, editorial page editor of the *Chicago Tribune:* "I don't see that there's a huge problem here. People aren't engaged in the process because there are no compelling issues driving them to participate. It would be different if we didn't have peace and prosperity."

ALTERNATIVES

Back in the sixties, when my friend Martha broke the news to her ol' daddy that she was going to leave her family home in the Greater Metropolis of Houston and head for the bright lights of New York City, he was not at all keen on the idea. "New York," he grumped, "why in the hell do you want to go to New York?" She told him that the city was full of culture and she wanted to experience it. "Culture?" he bellowed, "We've got culture right here in Houston." He paused for a second, then said, "Hell, we've got culture out the ass."

Finding the real news these days is a lot like Martha's youthful yearning for culture—sometimes you have to leave home to find it, especially if your "media home" is restricted to the networks and the conglomerate press. The good news is that good news *is* available—reliable, informative, unconstrained media sources that offer more than the minimum daily feed of officially sanctioned information bites, sources that actually tell you who is doing what to whom . . . and why.

The *Utne Reader,* the *Nation, Mother Jones, In These Times, National News Reporter,* and *Progressive* are only a handful of the dozens of popular periodicals on the national scene that routinely root out economic, political, and social stories that the establishment media ignore, belittle, or bury, and these periodicals are just the start of an honest, probing journalism that can also be found on the radio, in newsletters, via video, on the Internet, and elsewhere.

If you want to hear the viewpoint of someone besides Henry Kissinger about our government's corrupt China policy (Kissinger is a man who makes a bundle as a consultant to corporations doing business in China, by the way—a gross conflict of interest not generally disclosed by big-media interviewers who fawn over this shameless moneygrubber when he speaks in his role as a former secretary of state); if you're interested in knowing that local citizens are battling polluters in community after community, day after day—and winning; if you want to know what someone besides a pipe-smoking Wall Street economist or a boneheaded congressman thinks is needed for economic growth and prosperity; if you're curious about what the

CIA does between the times it's startled to find yet another double agent within its own headquarters building; if you wonder whatever happened to the farm crisis that was such hot TV a decade ago (did it just disappear when the cameras went away?); if you have heard whispers about the success of "living-wage" campaigns across the country, but cannot find them mentioned on the nightly news; if it would alarm you to learn that your tax dollars are financing the School of the Americas, a torture academy for Latin American dictators . . . indeed, if you want to know much of anything that's truly juicy in our country's economic and political doings, and if you want to get the information in a clear and timely fashion so you can act on it, do not count on "THE MEDIA." Instead, turn to "the media," those scrappy, sassy mags, rags, and wags steadily pounding out real news and inconvenient truths that inform you, even while they discomfort the privileged and powerful.

These range from long-established urban periodicals like the *Village Voice* in New York to upstarts like the *Progressive Populist* out of Iowa, from such venerable institutions as *Consumer Reports* to such fresh newsletters as the two-page *Rachel's Environment & Health Weekly,* from such national TV programming as *Planet Central* to the make-your-own programming of local "access" TV, from Pacifica Radio to AlterNet—and on into the boundless resources of the World Wide Web.

There even are a couple of mavericks dwelling inside the belly of the beast itself, still existing within the conglomerate media, though overwhelmed by the daily drone of corporate conformity that surrounds them. On the telly, for example, is C-SPAN, the marvelous invention of Brian Lamb, whose simple and subversive idea was to point a camera at official proceedings, click it on, and leave it there, letting us see the whole show without edits or "expert" commentary. C-SPAN has radicalized hordes of citizens who have, with their own eyes and ears, seen and heard the making of public policy, gradually recognizing that the system is largely in the hands of self-important, prattling fools, stooges for special interests—"Sweet Jesus," many a citizen has whispered, as reality unfolds on C-SPAN, "my senator has the IQ of a dust bunny."

A second bastion of journalistic integrity lurking within the confines of corporate media is the small space most newspapers set aside each day for the Last True Sons and Daughters of America's Revolutionary Pamphleteers: editorial cartoonists. I rank Tom Toles in Buffalo, Tom Tomorrow (AKA Dan Perkins) in San Francisco, Jules Feiffer in New York, Dan Wasserman in Boston, Nicole Hollander in Chicago, Herb Block in Washington, Ben Sargent in Austin, the duo of Gary Huck in Pittsburgh and Mike Konopaki in Madison, Ted Rall in New York, Signe Wilkinson in Philadelphia, Steve Brodner in San Francisco, Matt Wuerker in Portland, and other maestros of the 'toon craft as the best journalists in America, far ahead of any of the byline stars on the front pages. Not only do these cartoonists get the story right, going straight to the heart of it and skewering the bastards, but they even draw a picture of it for us!

Then there is the rambunctious bad boy of the media: talk radio. It has none of the intellectual cachet of print media, none of the sparkliness of TV, none of the hipness of the Internet—yet none of these combines the intimacy, accessibility, affability, passion, boisterousness, and pure plebeian punch of talk radio. Like a good corner bar, it is there and ready for you—you can get your regular shot of talk on commercial stations, on public and community stations, via short-wave sets, directly from the satellite, on the Internet (http://www.audionet.com) and even through what is called piebald syndrome (an actual physical phenomenon that occurs when a couple of your dental fillings next to each other act as miniature radio receivers).

The radio itself is so affordable and portable that it has become ubiquitous in our lives. You might not realize it, but if your family is typical you have at least five radios, not counting molars. Forty million Americans tune their sets to political talk shows with some regularity. I often hear progressive politicos complain that people won't come out to meetings anymore. Well, go where the people already are. Millions of folks wake up to talk every morning, drive to and from work with it, hang out with it at home, keep it on while at work (cabbies, nurses, mechanics, construction crews, farmers, restaurant workers, truck drivers, you name it)—and there even are radios you can take into the shower with you, which is more than I care to know about you.

Strangely, many liberals alternately fear and scoff at this lively medium, which they equate with Rush, G. Gordon, Ollie & Gang.

Yes, the medium presently is dominated by right-wing hosts, including noted criminals and kooks like Liddy and North, but (1) it does not have to be, and (2) this does not mean the listeners themselves are rabid rightists. Take Rush. Obviously the biggest bull in the business, but if you rendered the BS out of him he would fit in a matchbox. Talk-radio listeners largely know this. Surveys show that only about a third of his own audience agrees with him on most of his positions, about a third mostly disagree with him, and the other third are just shopping. The diehard Limbaugh "dittoheads," who have been the focus of much fear mongering in the media about rampant mindlessness among talk-radio regulars, amount to no more than 10 percent of his audience—about the same percentage of Americans who approved of Newt Gingrich's "Contract with America." Indeed, the recent slippage in Rush's radio ratings is the direct result of his identification both as Newt's national cheerleader and as Wall Street's chief apologist (it has been widely reported that talk listeners and callers tend to be antigovernment, but it is rarely noted that they are equally down on the corporate chieftains who are stomping on America's middle class, as Rush is learning the hard way).

The popularity of talk radio has nothing to do with right-wing hosts and everything to do with the simple fact that it is about talking. This is where ordinary folk hang out, hoping to get some answers, try out some ideas, shoot the breeze, mouth off, have a conversation, hear different (even strange) theories, laugh, cry, and enthusiastically kick the stuffing out of the establishment.

Larry Goodwyn, the eminent historian of American populism who now teaches at Duke, has expressed America's civic yearning in this touching phrase: "Our country is lonely." Lonely because grassroots people are ignored and dismissed, both politically and economically. People *want* to talk. But where? The political process makes a mockery of talking with people. In the last election, both Bill Clinton, with his totally scripted-for-TV bus caravans, and Bob Dole, in his hokey "Listening to America" sessions with carefully screened audiences, were interested in people only as stage props, not as real-life humans worthy of being heard. The major media outlets are even more aloof, talking down to people, not the slightest bit interested in what they are saying back. Ted Koppel even stoops to mocking

Americans' town meeting tradition, holding occasional electronic sessions that have no town and no meeting, featuring the same nine "experts" who are always chattering away on national television, and cutting off any impertinent (and, therefore, important) questions from the audience, which again is there strictly to serve as a prop.

No wonder so many people tune in to talk radio—it is about the last place where the vox populi matters. Rather than pointing fingers at this medium, liberals and progressives of every stripe should be embracing its audience and finding ways to *become* the medium. These are, after all, the "folks," most of them everyday working people who are mad as hell and ready to kick some serious butt. If we are not talking with them, then we cede the turf to the Limbaughs, whose schtick is to blame all the problems on "feminazis," "tree huggers," and "bedwetting liberals." Surely we are smart enough to counter such juvenile nonsense, aren't we?

In 1994 Bill Clinton called me for advice, which I freely gave, and he just as freely ignored. I have not heard from him since. I was hosting a weekend talk show on ABC Radio at the time, and he called me at my office to bemoan the daily tongue-lashing he was taking from the likes of Limbaugh, wondering how I thought he could combat it. By going on talk shows himself, I responded, pointing out that talk radio is not some hostile planet, but a place where he, with his gift of gab, would do well with the audience, no matter how big a jerk the host was.

Later I sent a note to my friend Harold Ickes, then a White House political operative who was close to Clinton, recommending that the President ditch that stilted five-minute radio message he broadcasts every Saturday morning (a format launched by the Gipper, who was a master at reading a script) and instead do a one-hour call-in show of his own, a format much more suitable to Clinton's ol' Southern-boy talky style. He would have been good at it, and what a stunning political coup it would have been, completely outflanking and discombobulating his right-wing radio berators by going directly to their audience and winning a bunch of them over. Plus he would have learned a lot—who knows, getting a weekly dose of real life from Betty in Baltimore and Tom in Tupelo might even have turned him into a Democrat and made him a halfway decent people's president.

But he did not. He did appear on a few talk shows, including mine once, but his major tactical approach to radio was to berate it, whining that it was shrill, mean, paranoiac, and dangerous, apparently oblivious to the fact that by chiding the medium he was also chiding its listeners . . . and losing them. So much for my influence in the White House. Shortly after my note, the Democratic Party began holding training sessions to teach activists how to discredit right-wing talk hosts. Of course, Limbaugh and others got copies of some of the training materials, howling, huffing, and hooting for weeks about the scaredy-cat Democrats. What a waste. Far better had the party held sessions to teach activists how to get their own talk shows.

Not that it is easy for a liberal, a green, a feminist, a populist, or any other brand of overt progressive to get on the air, but then it has never been easy for us, especially now that conglomerate ownership is sweeping up stations and networks wholesale.

I had my own experience with the stifling nature of media conglomeration in 1995 when I made an unscheduled trip to Disney World. My ABC weekend show had been airing for about a year, blasting the powers-that-be and preaching the populist gospel, when Disney Inc. announced on Tuesday, August 1, that it was buying my network. Suddenly I was the property of Mickey Mouse. On Friday of that same week the Senate passed the new telecommunications rip-off law, effectively opening a new and exclusive gold rush to a handful of communication conglomerates, authorizing a race to seize monopolistic control of America's mass-market news and information sources. On Saturday I went on the air and blasted both the Disney takeover and the Telecommunications Act, which Disney had lobbied furiously to pass. "I work for a rodent," is how I opened my Saturday show.

Turns out the mouse doesn't have much of a sense of humor, and there was an abrupt chilling in the network's enthusiasm for my program. Even though ABC had until then been committed to the steady growth of the show, even though we had built a bigger audience than the mighty Limbaugh had amassed at a comparable point in his show's development, even though we were adding sizable new markets at the time (Denver and Oklahoma City had just signed up in July, for example) and even though there were numerous advertisers available to back the show—I was suddenly moused, literally kicked off the air shortly after my anti-Mickey, antimerger broadcast.

There is an old gunfighter's tombstone in Arizona that reads: "I was expecting this, but not so soon." I have to concede that all of us involved in producing my show were rather astounded that hide-bound, totally corporate ABC had allowed a fire-breathing populist on the air to start with. Knowing that our broadcast could be short-lived, we had decided from the get-go to give the audience a full dose of populism, to speak the truth as we know it (which, by the way, is why I had to take on the Disney buyout of the network, intemperate as it is to piss on one's own hierarchy—but having pounded the podium regularly over other corporate power plays, I could not credibly have ignored this whopper). So we were hardly shocked to find our plug pulled, though it does stand as a chilling reminder of who is now in charge of the public's airwaves. Today's big-money owners are perfectly willing to have liberals on the air talking about welfare, Hillary's hairdos, abortion, Whitewater, prayer in the schools, and other social/cultural disagreements, but the one thing that Big Money does not want discussed is Big Money. Big surprise.

Periodically in our country, concentrations of economic, political, and media power become so great that democracy itself is stifled, and in such times it is the "alternative" media—the outsiders, the rebels and mutts of communication—that move to the front, connecting people, clarifying issues, giving perspective, pointing the way, focusing folks on action, speaking truth to power, and *making the difference.*

Thomas Paine and the other pamphleteers of the 1770s and 1780s were the alternative media of their time. Likewise, in the 1870s and 1880s, the populist movement of tenant farmers, urban laborers, and other grassroots forces found its way around the robber barons and the baronial press of that age by communicating through an alternative media of their own creation, including a network of more than 1,000 local newspapers (the Reform Press Association), such large-circulation magazines as the *National Economist* and the *American Nonconformist,* and a speakers' bureau of 41,000 "lecturers"—41,000 trained speakers who on any given night could spread out across the countryside and bring home the true news of the day.

So here we are, another hundred years down the road, facing an economic, political, and media concentration that is more powerful than ever, with Disney's muscular mouse and his conglomerate brethren having achieved a chokehold on our country's mass-market news machinery. But just as the pamphleteers and populists refused either to shut up or to be shut out, so is the spunky, innovative, entrepreneurial spirit of alternative media loose across the land again, already hacking new paths of communication around, over, under, and through the latest conglomerate media blockages. Truth will out.

My momma taught me long ago that two wrongs don't make a right, but I soon figured out that three left turns do. Within a year of getting kicked in the butt by Mickey's little yellow shoe, I was back on the air, thanks to the three-left-turn ingenuity of United Broadcasting Network. With ownership by a group of investors who are strong proponents of America's working families, including the United Auto Workers union, the very structure of this commercial network gives it a freedom of populist expression that none other has. Its programming—including my two-hour daily talk show—airs on hundreds of commercial stations coast to coast ("from Maine to Maui," as we enjoy pointing out), competing directly against Rush and the rest for the hearts and minds of the millions of listeners tuned in to mass-market radio.

What about advertisers, though? How can UBN rely on Exxon, GM, and other major sponsors to back a show like mine that routinely pops them upside the head for crushing workers, consumers, and the environment? The network doesn't. It does accept some product advertising, but mostly UBN is a radio marketer of made-in-the-U.S.A. products—like a home shopping channel on television. It buys the products directly from manufacturers and sells directly to listeners. Want a Regal "La Machine" food processor, a Creative Playthings swing set, some Oneida silverware or stainless flatware, a Hoover Steam Vac, a Roadmaster "Flexible Flyer" hardwood sled, a Rival "Crock-Pot," a Lionel train set, an Oster adjustable hair clipper, a Mountain Classic twenty-one-speed bike, a White Mountain six-quart ice cream freezer—all made in America? Call 800–888–9999 to get UBN's "Project USA" catalogue, which lists these and hundreds of other items. Not only are they all made by American workers, 70 percent of them are union-made, and all of them are price-competitive with what you find at the mall, putting the lie to the self-serving claims of such ethically challenged outfits as Nike that they *simply must* have their stuff made by nickel-an-hour workers offshore in order to compete.

By putting its made-in-the-U.S.A. principles to work as a marketing niche, UBN has found a way to finance its populist programming. Three left turns.

In 1988 I was to give a prime-time speech at the Democratic National Convention in Atlanta. I had carefully crafted my little gem of a talk and was ready for the moment, but I was nervous about the TelePrompTer, having never used one. These magical gizmos, used by Presidents and others for "Big Speeches," electronically display one's peroration in moving type on two glass panels in front of you, allowing you to read your speech without the audience knowing you are reading. To the TV viewers and the crowd in the hall, you appear to be a brilliant, extemporaneous orator who needs not a single note to expound so eloquently. Makes one wonder how William Jennings Bryan did without one of these things.

My concern, of course, was that I would be midway through my address, shining magnificently in the nationwide beam of the TV lights, and . . . boink . . . my electronic text would blink off. The convention planners, however, who insisted that all of us prime-timers use these speech-cheaters, assured me that they work perfectly and that I should not give it another thought—"Don't even bring your notes to the podium," they said, "we have fail-safe backup systems that guarantee the TelePrompTer will not crash." Well, they had run conventions and I had not, so who was I to question them?

The night before I was to have my big moment in the spotlight, Jim Wright, then the Speaker of the House and a terrific speechmaker, was on the podium having his moment—which did not seem to be going well. From my seat in the audience, I noticed that he was rambling and seemed to be telling old political war stories rather than delivering one of his usual well-polished jewels. I ran into him later that night and learned what had happened. He told me that things had started splendidly as he mounted the podium to the acclaim of thousands of cheering Democrats. He turned to one side of the great hall and waved, he turned to the other side and beamed, then he turned to the TelePrompTer to begin, and, sure enough, the speech was there and electronically ready for him. He began speaking the opening line, "When I was but a small boy back in West

Virginia . . . "—before he realized that it was Senator Robert Byrd's speech on the TelePrompTer.

Needless to say, when I got my turn at the podium, I had my handwritten notes tucked securely in my pocket. Never trust those in charge to put the right thing on the screen, on the air, or in print for you—always have alternatives.

4

POLLUTION

"STATUS QUO" IS LATIN FOR "THE MESS WE'RE IN"

THOSE DAMN ENVIRONMENTALISTS

They say if you get up in the morning and swallow a live toad, nothing worse will happen to you the rest of the day.

My toad for November 9, 1995, was this news item I found buried deep in the sports section of the *Austin American-Statesman:*

MERCURY ADVISORY

The Texas Department of Health has issued a warning concerning consumption of large-mouth bass and freshwater drum from several lakes. The warning also includes white bass and their hybrids from one lake.

The recommendation is that adults consume no more than two eight-ounce servings per month of drum or bass from Caddo Lake, Cypress Creek, Sam Rayburn, Toledo Bend and B. A. Steinhagen lakes. Children should eat no more than two four-ounce servings per month.

The white bass advisory is for Steinhagen Reservoir. The health department says adults should eat no more than eight ounces per month and children no more than four ounces.

Elevated levels of mercury were found in the listed fish after tests were run on a number of East Texas lakes. Officials can't pinpoint a specific source of the mercury.

Let's see now, was that four ounces of large-mouth from Caddo, or eight ounces of white bass from . . . where? What do they mean by "children"—under sixteen, under twelve, what? Mercury. That's a heavy metal right? Shouldn't this be on the front page? And what in darkest hell do they mean, "*Officials can't pinpoint a specific source of the mercury*"?

We are poisoning ourselves. More accurately, we are allowing a handful of profiteers to poison us. You do not have to be Rachel Carson to realize that letting this go on will suggest to the generation of the twenty-second century (if there is one) that the collective IQ of us twentieth-century Americans was a couple of digits dumber than beef jerky.

Still, the polluters, their protectors in Congress, their hirelings in big science, and their mouthpieces in the media join in a raucous chorus of scoffs every time anyone raises an objection, claiming any protest as nothing more than frivolous meddling by "those damn environmentalists," whom they caricature as a motley mix of professional worrywarts, half-batty bird watchers, and overaged flower children.

But hold your DDT-PCB-DBCP-CFC–2,4,5-T horses. "Those damn environmentalists" are not just a fanatic fringe—they are really "*us* damn environmentalists"—you, me, and Willie Nelson.

In 1987 Willie Nelson testified on my behalf at a hearing in the Texas legislature. I was state agriculture commissioner at the time, and one of my statutory duties was to regulate the use of pesticides, which—foolish me—I actually attempted to do. As Woodrow Wilson once said: "If you want to make enemies, try to change something."

Texas is one of the largest ag-chemical junkies in the country, consuming 40 million tons of herbicides, insecticides, and fungicides each year. This toxic stuff saturates our soil, runs off into our water supplies, gets onto our food and into our bodies. So in 1984 I began a program of curbing some of the grossest abuses—requiring, for example, that vulnerable farm workers and nearby neighbors be notified before any field was sprayed, and that they be informed about exactly what chemicals were being used.

Such a hullabaloo you never witnessed. The corporate ag establishment (led by the Chemical Council and its comic sidekick, the Farm Bureau) squawked like a banty rooster choking on a peach pit. What a racket they made, wailing that these "radical regulations" would outlaw farming as we know it and cause mass starvation in

Ethiopia, if not in El Paso. In every rural county they nailed my mug to the wall, put a political bounty on my scrawny hide, and howled that they would get me in my '86 reelection run.

They missed me that time, though, which only made them madder, more devious, and more determined. Enter Bill Clements, a Republican and a practicing doofus who was elected governor that year (during his campaign, he had tried to appeal to Mexican-American voters by letting it be known that he was taking Spanish lessons, which led me to blurt, "Oh good, now Bill will be bi-ignorant"). So, this peeved Republican was more than willing to help the ag establishment try to exterminate me through legislative action. They concocted a scheme in the '87 legislative session to move authority over pesticides from my department to another agency, where they could control it again.

Hence the Big Hearing. They had hoped to use their insider connections to muscle this change through the legislature quietly and quickly, but we took the ploy public, rallying ordinary Texans against it. From all over the state, people wrote, called, and even traveled to Austin to oppose any effort to dilute our tougher regulation of pesticides. They came not just from the Sierra Club but from such unexpected places as Republican clubs in Dallas. So many showed up for the hearing that it had to be moved to the floor of the House of Representatives, where we were ready with our surprise leadoff witness, Willie Nelson.

Willie, of course, is a renowned picker-singer-songwriter, but in Texas, at least, he is bigger than that—he's about as close as we get to true-life redneck heroes. His sixty-odd years have taken him along some mighty rough roads, and his hard-traveled, up-against-the-establishment, honky-tonking success is just the kind of thing to endear him to maverick Texans. To top it off, his Farm Aid benefit concerts have made him a genuine grassroots champion for the family farmer, so we had just the right guy going for us. His statement that morning was short and sweet:

> My name is Willie Nelson and I'm here to talk for our Agriculture Department.
>
> I'm here because I used to hunt rabbits up in Hill County—I understand we have some people from Hill

> County here today [cheers]. Me and my dog used to hunt rabbits around Abbott, Hillsboro, Itasca, West. Four or five years ago, I was up in Itasca visiting my cousin, Cecil Wilcox, and we were out walking around over the cotton fields. I said, "Cecil, where are all the rabbits?" He said, "I ain't seen any rabbits in years over here," and I said, "Why?" He said, "Pesticides." I said, "Well, if pesticides will kill rabbits, won't they kill people?"
>
> I was telling this to Bill Polk over in Dripping Springs, about the rabbits, and he said, "Well, have you noticed that we don't have any horny toads around either?" And I said, "Come to think of it, we don't, I haven't seen any in years." And he said, "Well, you know horny toads eat red ants, and when they run out of food they disappear, and we were killing red ants with pesticides."
>
> So I'm here today to represent the rabbits and the horny toads.

The committee, which included a bunch of toadies for the governor and backroom buddies of the chemical lobbyists, didn't move. For what seemed like ten minutes, they sat there all slack-jawed, staring at Willie, eyes blinking slowly—not knowing whether to question him, thank him, or just go bowling. One thing had dawned on all of them, though: Stopping pesticide regulations was not going to be as easy as the lobbyists had told them it would be.

After Willie did his number, after farm workers described being sent into fields while crops were still dripping with pesticides, after various farmers said *they* thought pesticide use needed more control, after some Republican mothers told about the afflictions their children had because of toxic chemicals—no member of the committee would even make a motion to support the governor's bill removing my pesticide authority.

The polluter establishment wants us to believe that environmentalism is the domain of elites, that the Great Unwashed majority is much too preoccupied with scraping together a living to spend time worrying about such prissy concerns as factory emissions and oil spills. Just get government the hell out of the way, they say, so developers can build, factories can roar, money can be made, and the marketplace can work its magic in solving both our economic *and* our environmental problems.

The purveyors of this conventional wisdom intentionally overlook the common sense of the ordinary American. There is a solid, unshakable and highly agitated environmental majority in our country—people who are not at all interested in some phony tradeoff between their families' environmental health and their economic health. They want unpolluted water, air, land, and food, period. This is a majority sentiment not only among the bean sprout eaters, but also among the snuff dippers of our country. Such folks are not likely to belong to the Audubon Society, most would not call themselves "environmentalists," and they probably know more about the PTA than the EPA—but they damn sure know pollution when it hits them. These are the "pollutees," and they are sick of the platitudes and pussyfooting they get from politicians when it comes to protecting their families.

- I'm talking about the family in the $60,000 cookie-cutter house in the farthest-out suburban development, right next to the cotton fields, noticing that their roses died after the sixth spraying this summer, remembering that the neighbor's cat died about the same time last year, and beginning to connect this to the recurring rash and cough their youngest child has.
- I'm talking about the working stiffs in the plastics plant who have learned that they can expect to die ten years earlier than normal because of where they work, at the same time they've learned that their company is cutting their health benefits, looting their pension, and lobbying to kill OSHA.
- I'm talking about conscientious mothers who carefully wash the fruit they give their children, but now find out that washing cannot eliminate the toxins and that the government's pesticide-residue standards fail to take into account that children's bodies are much more susceptible to chemical damage than are adult bodies.
- I'm talking about farm families, too, sitting at the kitchen table in the predawn light, talking about the herbicide they are about to spray, concerned by reports that traces of it are

now in their groundwater, and wondering if this stuff has anything to do with the tumor the doctor just found in dad's groin.

- I'm talking about folks who pack up the kids for a weekend getaway at the shore, the lake, the springs, or the river, only to find they cannot go in the water or eat any of the fish.

The vox populi is unambiguous, as poll after poll has shown year after year. A recent ABC News/Washington Post survey, for example, asked if people thought the federal government has gone "too far" or "not far enough" to protect the environment. Only 17 percent say "too far," and 70 percent say "not far enough."

Newt Gingrich's environmental wrecking crew has learned about the depths of this green sentiment the hard way. Fueled by the hubris of their 1994 electoral takeover of Congress and by a gusher of campaign money from polluters, the Republican-controlled 104th Congress locked arms with corporate lobbyists in a highly publicized assault not only on the laws and agencies, but also on the very idea of environmental protection. With exuberant cries of "Unfetter the Marketplace" and "Whack Big Government," they went after the budget and authority of the EPA and every other government office that stands between unbridled corporate whim and us. Apparently these gooberheads actually expected wild applause from a grateful public, but instead folks were widely appalled. After a year of the GOP's assault on the Environmental Protection Agency, for example, a Harris Poll found that 86 percent of us believe an aggressive EPA is needed today as much as or more than when it was created back in 1970. More stunning to Newt & Gang was that 76 percent of *Republicans* felt this way.

Meanwhile, in related polling news, public approval of the job being done by Newt's Congress had fallen further and faster than that of any Congress in history, and Newt's own approval rating was lower than that of mad cow disease, prompting his handlers to make him start wearing a paper bag over his head. Indeed, Republican pollster Linda DiVall reported in 1996 to her GOP clients in Congress that their party's environmental assault was disdained not only by Democrats but also by Republicans and independents. Her polling found that "55 percent of Republicans do not trust their party when it comes to protecting the environment," leading her to conclude, "Our party is out of sync with mainstream American opinion."

So did Newt back off? Nah. He's too tight with industry lobbyists and too hungry for industry campaign contributions to do that. Instead, party leaders advised their members to begin visiting zoos and engaging in tree plantings back home to bolster their environmental image.

I didn't want to be the one to break it to Newt and the GOP, but political tree plantings are as useless as putting lipstick on a pig—no one is fooled, and they are damn sure not going to kiss that pig.

WRESTLING THE WORLD FROM FOOLS

The Bible declares that on the sixth day God created man.

Right then and there, God should have demanded a damage deposit.

Included in the deal, says Genesis 1:28–30, was that man was to "rule over" the fish, the birds, the critters that move on the ground, the seed-bearing plants, and the whole kit and caboodle that Mother Nature has to offer. Now, "rule over" allows at least a couple of interpretations. First is the one celebrated in church sermons and on Earth Day, which is to consider ourselves the stewards of this bountiful paradise, accepting responsibility for the nurturing of things great and small. The other is the one actually practiced by those in charge, which is: "It's ours! Everything! God gave it to us. Grab it all, gobble it up, pave it over, make it pay or piss on it, get out of our way you enviro-weenie or we'll poison your dog."

The latest manifestation of this God-gave-it-to-me ethic is about the loopiest rationale for greed since the divine right of kings. It is called the "Takings Movement," which holds that if the public and its government do not allow a property owner to pollute at will, then the same public and its government must pay said property owner cash money not to pollute. Say what? Yes, they insist, the Fifth Amendment to the U.S. Constitution means that if Big Belcher International acquires property next to you and is subsequently prevented by government regulation from building a toxic waste incinerator there, this government action amounts to a "taking" of BBI's private property, and we taxpayers owe them for the money they would have made incinerating toxic waste.

If the Mafia tried to run a racket like this it would be the subject of an FBI strike force, seven or eight grand jury investigations, and a whole new round of *Godfather* movies. But since the takings racket is being run by a higher class of mob, with proper Wall Street addresses, it has already been sanctioned by law in twenty states and is one of the prizes Newt Gingrich slipped inside his "Contract with America" for his corporate contributors. As Andy Young, former mayor of Atlanta, once stated, "Nothing is illegal if a hundred businessmen decide to do it."

Of course, the real takings are not being done by the government, but by the same Big Belcher Internationals that are pushing this sophistry. The good news is that in the face of such polluter arrogance in our national and state capitals, ordinary folks have begun doing their own "takings"—taking charge of their environmental destinies by taking on the polluters at the community level. As Patti Smith sings in her hard-rocking anthem: "The people have the power/The power to dream, to rule/To wrestle the world from fools."

Sure, you say, nice sentiment, but show me on what planet this is true.

Planet Earth, for one—not the whole planet, obviously, but in thousands of locales around the globe, where fed-up locals are grabbing down-and-dirty speculators and spoilers by the short hairs and turning them every which way but loose. In our country this is not being done by groups that the TV networks or Washington politicians or even the big brand-name environmental organizations pay any attention to, yet they are, collectively, making the most significant progress against polluters. They do not have multimillion-dollar budgets, do not publish beautiful photo calendars, and do not have professional lobbyists. They are much scruffier, scrappier, and passionate, with in-your-face names like JAWS, GASP, WOW, SOS, WARN, SOCK, and WAR. These are groups that pop up when something happens that causes hardworking, perfectly peaceful, average citizens to get hot as hell and spontaneously combust.

Mary Hernandez is one such. A thirty-five-year-old insurance auditor and mother of three, she lives in a working-class neighborhood on the east side of my town, Austin, Texas. A proud mom, she often bragged about her son Antonio growing up to be President. But then he began to get sick a lot, and she began to wonder if he would grow up at all. Talking with neighbors, Ms. Hernandez learned that a lot of people in her area, especially children, were chronically sick with such ailments as nosebleeds, headaches, nausea, dizziness, asthma, and rashes.

Odd. The neighbors told others and began comparing notes, looking for a pattern to their illnesses. Then they realized the one thing they had in common was right in front of them: All lived within a few blocks of a fifty-two-acre tank farm where Exxon, Chevron, Citgo, Texaco, Mobil, and Coastal States owned and operated forty-three massive storage tanks holding 10 million gallons of gasoline. Eighty percent of the gasoline used by Austin motorists sat in their little neighborhood.

About this time, another east Austinite, Sylvia Herrera—a mother of two children who was in graduate school at the University of Texas and had done research on the health problems of people exposed to industrial toxins—came across an item tucked quietly in the classified section of Austin's daily newspaper giving "public notice" that Mobil was seeking a state permit to expand its tank-farm operation and *to continue emitting toxic chemicals into the air.* To continue what?! This was the first time the people had any knowledge that poisons were being released right in their midst.

Herrera, Hernandez, and dozens of other increasingly agitated neighbors promptly joined forces in a new grassroots group called PODER. The acronym stands for People Organized in Defense of the Earth and its Resources, but the name gets attention in the community because *poder* is the Spanish word for "power," and that is what these people were after—the power to stop the poisons. Not only were their own homes exposed, but so were three small groceries where they bought food for their families, and so were two elementary schools—one within 400 feet of the tank farm and another within 1,000 feet.

First they gathered information, going door-to-door to build a more thorough record of illnesses and going to libraries and agencies to learn what was coming out of these forty-three storage tanks. "I didn't even know what benzene was or what it could cause, but I sure do now," Mary Hernandez says. What she learned is that it can cause leukemia, among other cancers, and that some of the early symptoms of benzene poisoning are headaches, dizziness, and nausea—precisely what her son was experiencing.

Then they learned something that is always somewhat shocking to people, even in these cynical times: The government agencies they assumed were protecting them, were not. Records at the Air Control Board revealed that state, city, and county health officials had known about the benzene problem for some time and were "concerned," but never mentioned it to the neighborhood. A writer for the *Polemicist,* an Austin alternative newspaper, dug into files at the state Water Commission and found that there had been a history of spills and leaks at the tank farm, but no one in the neighborhood was ever notified. The agencies knew, the oil corporations knew, everyone knew except those who most needed to know—the people living next door.

The members of PODER set out to take their government back from the polluters and make it work for them. They held weekly organizing meetings at the neighborhood library; they got a resident from each street to serve as block representative to get information out quickly and to mobilize people; they enlisted an African-American community coalition called EAST to join the fight; the principal of the area high school demanded a meeting with Texaco; the PTA petitioned the state agencies to test for contamination of the air, water, and soil around the schools; residents started writing letters to agencies and elected officials—suddenly the neighborhood made itself a blip on officialdom's radar screen.

Then, in one stroke, the blip became a force to be reckoned with. PODER invited elected officials and the media to take a "Toxic Tour" of their neighborhood. This put the tank farm and PODER on the political map, and things began to happen in a hurry:

- **March.** Prodded by the publicity and by suddenly awakened politicians, state agencies that had been curtly dismissing the neighborhood's complaints and had earlier denied there even was a problem, begin testing and quickly find that (well, my, my!) the spills and leaks have indeed spread in plumes from beneath the tank farm into the neighborhood, creating pools of poison that contaminate yards, creeks, and wells.
- **April.** The Water Commission orders the companies to clean up the contamination. The Air Control Board finds numerous violations, including nonfunctioning pollution-control equipment. County tax officials devalue the worth of all homes within a quarter of a mile of the tank farms by as much as 60 percent. Homeowners sue the oil companies for, in essence, "taking" their property values from them.
- **May.** The Department of Health finds that 68 percent of the families surveyed in the neighborhood reported health problems consistent with exposure to toxins. The *Austin*

American-Statesman reviews county medical records and finds that 64 percent of all bronchial pneumonia cases come from this one neighborhood, as do between a quarter and a half of various other respiratory and stomach disorders. Escalating the fight dramatically, County Attorney Ken Oden launches a special investigation into criminal and civil wrongdoing by the oil companies.

- **June.** In an unprecedented agreement with the Air Control Board, the companies agree to reduce toxic emissions by two-thirds. The county's top health official declares the tank farm an "unacceptable health risk" and says it should be removed.
- **August.** Chevron announces it will close its terminal. In a PR stunt, Chevron offers to donate its building at the site to the community for a health clinic. PODER says thanks but no thanks to a health clinic on property soaked in poison.
- **September.** Mobil, Texaco, and Coastal States say they will leave, too. County Attorney Oden intensifies his investigation, and the grand jury subpoenas Exxon documents.
- **October.** Citgo announces it is out of there, leaving only Exxon. Oden sues Exxon for pollution violations.
- **November.** PODER and other neighborhood groups launch a citywide boycott of Exxon stations and products.
- **January.** Mobil and Coastal States are the first to keep their word, announcing that they have closed their facilities permanently and will proceed to dismantle their tanks and pipelines, as well as begin cleanup work. Exxon says it has no plans to move and is committed to "continuing to operate as we have in the past." The consumer boycott of Exxon continues, as does Oden's investigation.
- **February.** Neighborhood residents, joined by Austin environmental activists, announce plans to travel to Houston to picket Exxon headquarters. Four days later Exxon announces that in the face of "prevailing negative sentiment in the Austin community and the likelihood of lengthy and costly litigation," it, too, will close its fuel-terminal operations in the neighborhood and pay a $19,000 fine for air pollution violations.

Susana Almanza, cochair of PODER, said of this year-long fight: "People have learned that they may not be rich, they may not be politically connected, but that they do have a voice—and they can make things happen."

She is right. Environmentalism is not just about protecting our environment, but also about exercising our power. It is about Mary Hernandez joining with her neighbors in PODER and daring to confront Exxon, Chevron, and the other corporate powerhouses that, aided and abetted by our own government, are taking from us, and taking and taking and taking. The takings will not stop until we stop them. As Mary Hernandez said after helping wrestle her small piece of the world from fools: "Our kids were sick. Somebody had to do something."

NIMBY

Sometimes patriotism asks too much, even of such flag-waving, proud-to-be-American patriots as the corn, cattle, and cotton farmers on the High Plains of the Texas Panhandle.

Farmers tend to be optimists in a pessimistic business, and more often than not they will hunt for the bright side of most anything that is thrown at them. My Aunt Eula, who farmed, was like this—if three bales of hay fell out of the loft and knocked her flat, she'd stagger back up, dust herself off, and say, "Well, it could've been four." In the Panhandle, this yeomanly optimism is reflected in the names of their towns—names like Sunnyside, Dawn, and Happy (Official Motto: "The Town Without a Frown").

One bright winter day back in the eighties, though, an announcement was made at an Amarillo press conference that did bring a frown to Happy and to just about all the weathered faces of the Panhandle's hardworking people. They watched that evening's news in dismay as federal officials, standing with beaming representatives of Amarillo's chamber of commerce, revealed that the lucky locals were to be the beneficiaries of a multibillion-dollar, job-creating, pride-instilling, red-white-and-blue federal construction project. After a nationwide search, neighboring Deaf Smith County had been deemed the very best place there could be to serve as the "high-level nuclear waste repository" for the entire United States of America, forever and ever, amen.

It was a simple matter of geology and technology, the feds explained, noting that some splendid salt domes about a mile beneath Deaf Smith County's rich soil would be perfect for storing waste from Three Mile Island, Rocky Flats nuclear weapons plant, and every other atomic waste–producing source in the country. The engineers would simply bore some wide shafts down to the domes, carve out some storage space, and then, for the foreseeable future, trainload after trainload of nuclear nasties would arrive daily and be deposited. Perfectly safe. Trust us.

Sure, thought the farmers, about like a chicken trusts Colonel Sanders. To get to the salt domes, the shafts would have to be drilled through the Ogallala Aquifer—the chief source of water for the whole Panhandle and the largest freshwater aquifer in the United States, spanning eight states. Did these brilliant scientists know about the geological fissures, faults, and seismic activity in the area? (Turns out they did not.) What market in the world was going to want to buy corn, cotton, and cattle raised above a nuclear dump? Who wants to live atop a nuclear dump? Do they think everyone out here just fell off a turnip truck?

At a packed meeting in the courthouse in Hereford, a federal man attempted to dismiss the people's concerns as so much hysterical know-nothingism, asserting calmly that nuclear engineers and public officials far more knowledgeable than they had approved of the scheme, adding that Amarillo's business leaders and daily newspaper backed it fully, and commenting that he was certain that the great majority of Panhandle people were for it, too—after all, *America!* needed this repository, and the people of this area would surely want to do their patriotic duty, as they had done so often before.

"*Stop right there,*" thundered a voice from midway back in the crowd. It was Carl King, a corn farmer and the kind of guy who flies the Stars and Stripes on his tractor, his pickup, his house, and his gimmie cap. "Patriotic don't mean being stupid," Carl bellowed, his entire bald head flushing red with emotion. "We don't care if you put that crap in the Statue of Liberty and try to send it out here, WE DON'T WANT IT!"

Raucous applause and huzzahs for Carl. The fed fellow didn't realize it, but the deal was dead right then and there. It took a few more years, a lot of citizen agitation, and a couple more votes in the Congress, but there is no nuclear waste repository in the salt domes of Deaf Smith County today.

Industry and government do not like people like Carl King. "NIMBYs," they snarl at Carl and any of us who dare to stand up and say: Not In My Back Yard. Like *we* are the ones being irrational here.

We are told that the NIMBY attitude is antiprogress, antiscience, and anticivic. Obstructionism for obstructionism's sake. Whether they are trying to locate a waste dump, a lead smelter, a landfill, a plastics plant, a toxic incinerator, you name it—they tell us that it has to go somewhere, that not everyone can be a NIMBY, and that if it is not to go in our neighborhood or town, then exactly where do we suggest they put it?

It is really tempting to tell them, isn't it?

One place you can be certain it will *not* be placed is in *their* neighborhoods. You can drive all day long through Beverly Hills, River Oaks, Nob Hill, North Shore, Grosse Pointe, White Plains, West Palm Beach, Highland Park, or any of the other enclaves of the wealthy and not see a single pit, smokestack, terminal, or other sign of "progress." If NIMBY is so bad, why are offending facilities Never In Their Back Yards? NITBY.

Pollution is a class issue. If it spews, burbles, oozes, blasts, emits, gushes, radiates, or otherwise does something unpleasant and dangerous, it is located in working-class, rural, minority, or low-income communities—usually in places with some combination of the four, places with little political or economic clout. If you are poor, your chances of having Mr. Pollution as a close neighbor are one in six. Minorities are especially hard-hit—60 percent of African-American and Hispanic-American families live in communities with at least one toxic waste site.

And—in an Amazing Coincidence that never fails to baffle authorities—such neighborhoods also tend to be the ones that show up on maps depicting cancer clusters, concentrations of birth defects, and other environmental health problems.

In today's world of arrogant and willful polluters, compliant government, and vapid media coverage, what they derisively refer to as the "NIMBY Syndrome" is far from irrational—it is the only sane response. The problem is not the people, it is the insanity. Why do we continue to make all these poisons if there is no safe place to make them, and no safe place to store the waste? Nuclear power, for example, is not a necessity to anyone but the grossly subsidized and heavily indebted nuclear power industry. There is no shortage of alternative energy strategies (including conservation) that are cheaper, safer, and generate no toxic waste problems for us or our progeny.

The old vaudeville joke teaches us something here: A guy goes to the doctor and says, "Doc, it hurts when I do this," stretching his arm behind him in an unnatural motion. So the doctor says, "Well, don't do that!" It is a stupid joke, but not as stupid as nuclear waste. NIMBY is simply nature's way of saying, "Don't do that."

SCIENCE

When I was a barefoot nubbin of a boy growing up in the early fifties, one of my most joyous summertime moments was when the town sprayer came chugging down our alley, spewing out a great, billowing fog of DDT. From two or three blocks away, other kids would sound the gleeful alert—"Sprayer's coming, sprayer's coming!"—signaling my brother and me to bolt from the house and join them. A whole gaggle of us would then run whooping and hollering right along behind the sprayer, blissfully enveloped in the mist, gulping lungsful of the stuff as though it were laughing gas, instead of poison.

No one mentioned to us, our parents, or even to the guy driving the sprayer that this mosquito spray had even a whiff of danger to it, unless of course you were a mosquito. Quite the opposite, we were flat-out told by the high priests of industry, government, and science that DDT was our friend, another miracle from the beakers of modern science. We trusted them. The fifties were like that—*Ozzie and Harriet* and an unblinking trust in America's scientific beneficence, a time when we eagerly believed in the broadly advertised promise of "Better living through chemistry." So any raised eyebrow by a neighbor busybody about allowing kids to play in the spray would be met with "Aw, it can't hurt the little squirts. And I haven't seen a mosquito bite on 'em in days."

Since that innocent time, hundreds of thousands have learned the hard way that prolonged exposure to this organochlorine interferes with human hormones and causes birth defects, reproductive maladies, and cancer, especially breast cancer, which now kills 50,000 women a year. One of the highest rates of breast cancer in our country today, for example, is found on Long Island, New York, where potato farms were regularly and heavily doused with DDT in the fifties. This chemical has since been banned in the United States, but there is no escaping it. It takes hundreds of years for DDT to break down in nature. So it lingers still in our yards, riverbeds, lakes, fish, birds—and in us.

The lesson here is that science is not truth, and it often errs. The one word you never want to hear your dentist say while elbow-deep in your mouth with sharp tools and screaming drills is: "Oops." But that is exactly what happened with DDT—the "oops factor." This is not something that Established Science or those who profit from it want to discuss, but it affects us at least as profoundly as do the successes of scientists. From DES to IUDs, from Agent Orange to nicotine, from Love Canal to Silicon Valley—what was touted as scientific truth only yesterday turns out to be wrong today. Oops. Sorry if this caused you any discomfort or death.

As Texas commissioner of agriculture for eight years, I regularly encountered Ph.D.-flaunting scientists from ag colleges and industry who absolutely insisted that each of the thousands of insecticides, herbicides, and fungicides on the market was there because humankind could not survive without it and because each one had been so rigorously tested for human safety that they would gladly swig a cocktail of pesticides in front of the Alamo at high noon. It was a level of true belief rarely seen outside cults, and I finally came to the conclusion that they had, indeed, been sniffing too much of the stuff in the lab. Nothing else could explain such detachment from reality—a reality that includes this frightening saga: Way back in 1972, Congress instructed EPA to reassess the safety of 590 chemical compounds that are the active ingredients in some 18,000 pesticide products. It had been discovered that many of these compounds had been approved for use on the basis of bad data, so Congress ordered a retesting program, giving our environmental protectors five years to do this urgent job. But they ran a little behind . . . and they still are. So far, twenty-five years later, EPA has completed the reassessment of only 152 of the ingredients. This leaves 438 of them untested, but still on the market, still being put on crops, still drifting through the air and running off into streams, still working their way up the food chain . . . into us.

There is one newspaper column I always try to catch. It is a daily itemization of ten or so curious facts, quotes, oddities, and tidbits that are assembled and wonderfully presented by L. M. Boyd, a wily wordsmith based in Seattle. With some of his more astounding "facts," though, Boyd attaches a clever qualifier: "Interesting if true." No more credulity than this should ever be given to the pronouncements of Established Science, especially when it is stepping forward to sanction some particular profiteering scheme of natural resource plunderers, poisonous product manufacturers, or toxic waste handlers. As one fully credentialed but honest scientist confides: "Putting an awful lot of stock in our theories is not a good idea. Science is about being wrong."

Yet those who make environmental policy in Washington and most state capitals today are out to deify "science" and to give certain of its bishops a privileged position at the policy table. In Newt Gingrich's Congress, there is a feverish ideological crusade among both Republican and Democratic Party conservatives to install panels of scientists at the top levels of environmental and public health agencies, and to route all policy decisions through them. Should the EPA license a toxic waste incinerator in Columbus, Ohio, right in the midst of homes and schools? "We can't have environmental hysterics drive a big economic decision like this," the polluters' politicians say. "Such decisions should be above politics. Let the white-smock boys handle it in an unbiased and factual, scientific sort of way."

"Unbiased and factual" on whose side? On industry's side, of course, since practically all the scientists these lawmakers seek to impanel are pro-whatever-industry-wants. For example, attached to the '96 Farm Bill is an amendment creating the "Safe Meat and Poultry Inspection Panel." Composed of seven scientists, the panel was given broad powers to review decisions of the U.S. Department of Agriculture's Food Safety and Inspection Service.

The department had just promulgated some new regulations requiring industry to modernize its antiquated method of protecting us consumers from tainted meat. Believe it or not, the method in use amounted to nothing more sophisticated than "poke and sniff"—some geezer would poke a finger in each batch of hamburger or into the carcass of a chicken, sniff his finger, and declare: "Nope, 'tain't tainted." About 5 million of us consumers get sick and more than 4,000 die each year from eating burgers, chicken, and other meat that supposedly "'tain't tainted," learning the hard way that the old techniques do not detect such microbial contaminants as *E. coli* 0157:H7 bacteria. This is the exotic, virulent, and totally nasty pathogen that made more than 500 people in Seattle convulsively and horribly ill in 1993, killing three children, after all had chowed down on hamburgers at a Jack in the Box fast-food joint.

There was such a roar of public outrage over this *E. coli* 0157:H7 outbreak that even the Department of Agriculture, the long-kept mistress of the meat industry, was forced to take some reluctant steps toward reform. Not right away, of course—it took USDA three years and about a thousand more *E. coli* deaths to produce its new set of sanitation and testing rules for meat processors. But at long last, in the spring of '96, the modernized standards were ready to be implemented and (O, Progress!) to stop our own hamburgers from killing us.

But: "Halt!" "Stop it right there!" "Do not take another step!" At the behest of some industry powerhouses, here came a clown car filled with congressional ninnies waving their oversized monkey wrench called the Safe Meat and Poultry Inspection Panel. With SMAPIP tossed into the regulatory mechanism, the new microbial testing requirements were held up for several more months, and all future USDA food safety rules have to be funneled into this new bureaucratic meat grinder.

Delay meant more deaths—the statistics were clear on this. Yet congressional leaders trumped this concern with a statistic of their own: 41 million. That is the number of dollars put into congressional campaign coffers by food corporations during the decade leading up to the SMAPIP amendment. This cash, plus millions more to hire teams of Washington lobbyists, has created what the Center for Public Integrity, a watchdog group, calls one of Washington's "most effective influence machines," and it has prevented any significant food safety legislation from coming out of either the House or Senate agriculture committee in this ten-year period. Among the meat processors fueling the influence machine with more than half a million bucks each in campaign contributions are Cargill Inc., Tyson Foods, and ConAgra Inc., and among the big recipients of meat industry money is the head of the House himself, Newt Gingrich, who swallowed a quarter of a million big ones. Lawmakers got their money, and industry got its SMAPIP.

Good luck with your hamburger, because what Congress did by creating this "scientific" panel was to give Tyson, IBP, Hormel, Jack in the Box, and other corporate meat purveyors official control of the process. Their amendment explicitly allows scientists *paid by the meat industry* to serve—indeed, five of the seven members of the panel *must* be animal-health scientists rather than human-health experts. In the first place, this is the equivalent of putting a veterinarian in charge of your child's health. Far more dangerous, though, is that animal-health scientists are notoriously subservient to the very industry from which we are trying to protect ourselves, the industry responsible for so many *E. coli*, salmonella, and other deadly outbreaks from tainted meat.

It will come as no surprise to you, then, that the Clinton Administration's final meat inspection rules were far less bold than the Ag Department's press release proclaimed them to be on July 6, 1996. "Meat Inspections To Be Tougher," headlines happily exclaimed the next day, but "tougher" turns out to be so squishy that the industry can continue to produce meat with the exact same level of contamination it contained before the rules. For example, prior to July 6, 1996, one out of every five chickens and one out of every two turkeys on the market were contaminated with salmonella. After July 6 one out of five chickens and one out of every two turkeys *still* come with salmonella. The only difference is that the industry's bad-bird rate is now officially sanctioned in law. And while "poke and sniff" testing has indeed been replaced by high-tech, microbial analysis that can detect *E. coli* 0157:H7, you still might want to char the hell out of your burgers, because USDA's "tough new regulations" provide that *the industry itself*, not government inspectors, will do the testing.

Panels like SMAPIP have become shams, lending an aura of scientific objectivity when they actually are little more than legalized mechanisms for installing foxes in henhouses. Corporations can brashly put their own in-house scientists on these panels, or they can be slightly more foxy by having university-based scientists take the seats at our government's policy table. Do not be snookered by the appearance of independence bestowed by the cap and gown of these academicians, though—even scientists employed at our public universities commonly work under research grants awarded by food processors, moonlight as industry consultants, and are paid to sit on corporate boards.

Just like politics, too much of modern science follows corporate money. Jobs, contracts, honoraria, prizes, publications, and other awards are carefully parceled out to Ph.D.'s who walk the line. Not only does this growing dependency tarnish the integrity of science, but it also corporatizes our nation's scientific agenda. For example, with so much public concern about toxic chemicals messing with our cells, genes, and hormones, why does it receive so little scientific inquiry? Consider breast cancer. Thousands of scientists in companies, universities, foundations, and government agencies are studying the hereditary and dietary causes of this epidemic disease, yet it is conceded that as few as 30 percent of breast-cancer cases can be traced to these causes. The other 70 percent of cases are widely suspected to be caused by manmade poisons in our environment, yet there are not a handful of scientists looking into this.

Why?

Because practically no money and even less professional prestige is awarded to anyone who wanders off the well-beaten academic path into this wilderness, and because a hostile corporate and political hierarchy lurks out there, eager to devour juicy adventurers who dare suggest a link between chemicals and disease. If you are a scientist with enough smarts to spell C-A-R-E-E-R A-D-V-A-N-C-E-M-E-N-T, you look at a potential study of, say, organochlorines and birth defects, and you ask yourself: "Will Monsanto fund this? What journal will publish my findings? Who will pay an honorarium to present a paper on it? Will it lead to a promotion, a consultancy, what? Hmmmm, why bother?"

Then there is the "we are science and you are not" school of elitism so deeply embedded in Establishment Science. This charming attitude shuts down even fully credentialed scientists who happen to have a stray, unconventional idea, and it shuts out the uncredentialed mavericks who just might be onto something. Throughout my tenure as Texas agriculture commissioner I encouraged the mavericks, largely because they seemed to have ideas that actually worked, which distinguished them markedly from the chemically addicted Ph.D.'s of aggieland and the innovation-impaired agents of the government's agricultural extension service. Better to be degreeless than clueless, it seemed to me.

As Ralph Waldo Emerson once noted, common sense is "genius dressed in its working clothes," and I came across a lot of genius out in the countryside, every bit of which was having to butt heads with the official order. Buddy Maedgen, a third-generation farmer near Eagle Lake, Texas, was one of these commonsense geniuses. He and his family had farmed conventionally for years, until the mid-eighties when they began to notice some alarming side effects to the pesticides they were putting on their crops: calves dying, chemical drift killing other crops, a farm worker's family getting sick. It led Buddy to begin experimenting with nontoxic pest control methods and ultimately to build one of our country's few insectaries—a facility for raising insects that are natural predators of the bugs that are agricultural pests. The idea is simply to let the bugs duke it out—*bugo a bugo*.

For example, he bred warehouse pirate bugs that could be released in grain elevators to kill the insects that infested and ate the grain. The warehouse pirate bug has a very brief lifespan and dies shortly after doing its helpful work; its remains are then easily sifted out before the grain goes to the mill. The entire process is cheaper, safer, and more ecologically sound than the standard process of periodically dousing the stored grain with chemicals to kill the grain-eating pests.

Buddy found eager customers . . . until the bug police stepped in to stop him. The Food and Drug Administration declared that adding insects to a grain elevator amounted to "food adulteration" under federal rules, even if the beneficial bugs do sift out in the process. Adding chemicals however, was not adulteration according to the feds, because science had established a "residue tolerance level" for the chemicals—an amount of pesticides that the scientists *promise* is safe for us humans to swallow. Never mind that the government's pesticide-residue data range from flawed to falsified, here we have presented to us a fine sample of the unnatural logic of official science: Better to have a pest control system that leaves you eating trace amounts of pesticides than one that "just says no" to the chemical makers and gives you perfectly uncontaminated grain.

But the FDA had spoken, and suddenly Buddy's bugs were outlaws. That would have been that, another solid victory for "science" and the insecticide industry it was fronting, except for one thing: People balked. Buddy's customers wrote letters and made calls. My office began rattling some cages, and a couple of U.S. senators got on the FDA's case, too. Then CBS's *60 Minutes* put a fifteen-minute spotlight on the bug brouhaha, which caused a national uproar of pro-Buddy political support. Finally, after a four-year fight, enough public pressure was applied that FDA had to back off its silly scientific posturing and succumb to common sense. At last Buddy's bugs had a federal license to do what they do naturally.

This was but a hiccup, however, in industry's determined campaign to install their scientists at the wedge point of environmental decision making. In setting policy, intoned former EPA administrator William Reilly, "Sound science is our most reliable compass." Nonsense. Even if he meant truly independent science, as opposed to corporatized science, which he did not, sound science is our most reliable science, nothing more. Scientific findings are certainly a legitimate consideration in dealing with pollution, but so are political findings and, for that matter, so are culture, folk wisdom, and good old-fashioned American suspicion about what the powers-that-be are trying to do to us.

Rita Carlson is no scientist. She is a homemaker and an environmental hellraiser. She did not want to be a hellraiser, but pollution has made her one. Until recently she lived twelve blocks from the Union Carbide plant in Texas City, located on Galveston Bay at the Gulf of Mexico. Each week 200,000 pounds of toxic chemicals spew from this massive plant. There are eight such chemical factories in Texas City, and twenty-nine others in neighboring communities along the bay. Nearly a third of the nation's chemical products—from plastics to pesticides—come from these twenty-nine plants.

Alarmed by illnesses in her family, Carlson began her own investigation and found a cancer hot spot in Texas City. On one street of twenty-two homes, seventeen families have cancer. In some houses, both husband and wife have cancer. Cats die of cancer here. Children have kidney transplants and others have holes in their sinuses eroded from breathing the polluted air. Carlson has a tumor in her sinuses and her son has abnormal lymph glands.

She helped organize the community to fight for pollution controls, earning the enmity of both industrial and public officials, who branded her a hysterical agitator. The fight goes on, although Carlson herself has had to move away to try to save her family's health. "They say we are only screaming women," she comments. "I think the reason you see a lot of women more involved in environmental issues is because it's the women who have to rock the sick babies at night."

HOGS, TURKEYS, AND MAD COWS

Gib Lewis, a former speaker of the Texas House whose tenure primarily will be remembered for his endless ability to mangle the mother tongue, once observed during debate on a bill: "There's a lot of uncertainty that is not clear in my mind." If you knew Gib, you would know how true that was.

In fairness, though, the Gibber is not the only one unclear about actions of the Texas legislature—a place that former member Bob Eckhardt described as "Built for giants, inhabited by Pygmies." (Under pressure, Bob later apologized. To the Pygmies.) But even the most hardened watchers of the biennial legislative mayhem practiced by this menagerie were stunned in the '95 session when a majority of members rose up on their hind legs and voted aye to a bill making it an actual crime to disparage food.

Isn't it good to know that murder, rape, hate crimes, and other heinous acts are a thing of the past so that, at last, our legislators can focus on such pressing criminal activity as . . . food disparagement?

Are you guilty yourself? C'mon buster, confess: I bet you have badmouthed Brussels sprouts in your time, or maybe you have ridiculed raddichio, cast aspersions on asparagus, or pooh-poohed prunes.

OK, it is not the idle maligner at the dinner table that the food industry is out to gag, but the journalists, the consumer and environmental groups, individual pure-food advocates, and anyone else who dares to raise public questions about the safety of the industry's products and processes. Say a mad-as-hell group called MOM (Mothers Offended Mightily) springs up in East Jesus, Missouri, to protest some industry practice that they fear is doing harm to their children—residues of methyl bromide on strawberries, synthetic bovine growth hormone in milk, irradiated bananas, *E. coli* bacteria in frozen hamburger patties, just to name a few of the real-life risks families face in the modern meal.

Already industry lobbyists in Texas, Ohio, Alabama, Arizona, Idaho, Colorado, Florida, Louisiana, Mississippi, Oklahoma, South Dakota, and Georgia have rammed through disparagement laws that would let corporate lawyers sue MOM and haul these ladies into state court for doing "harm" to the reputation of the industry's edibles. Instead of the burden of proof being on food manufacturers to show that their chemical, genetic, and other tampering with our food supply is safe, these ludicrous laws put the burden on MOM to prove beyond a shadow of scientific doubt that methyl bromide, synthetic BGH, and all the rest are *un*safe. It is a form of legalistic intimidation that says to MOM: Either spend hundreds of thousands of dollars on lawyers and experts and endure a lengthy trial in which the rules are stacked against you . . . or shut up and eat your dinner. Even though four out of five Americans today tell pollsters they are "*very* concerned about food safety," these food disparagement laws free companies to do their damnedest to our dinner, then defy us even to question their madness.

One who did defy their madness, challenge their dippy disparagement laws, and pretty much rub their noses in their own mess was someone they stupidly picked on themselves: the Queen of Daytime TV, Oprah Winfrey!

On her talk show in April 1996, a guest noted that Oprah's viewers would gag if they knew that America's corporate beef barons were feeding a disgusting mix of rendered cow parts to cattle, adding that this ration was creating a form of mad cow disease in our food supply. A startled Oprah declared that this news flash "just stopped me cold from eating another burger: I'm stopped."

That day, cattle prices on the Chicago Mercantile Exchange took a 2½ percent tumble, and next thing you know—yippie-ti-yi-yo!—here came a whole posse of corporate cowboys shooting from the lip and demanding Oprah's hide! Led by Cactus Feeders Inc., a Las Vegas–based corporation that is among the world's largest feedlot operators, these boneheaded buckaroos thought they had Ms. Winfrey dead to rights on a disparagement violation. So off they galloped to the courthouse in Amarillo, Texas, claiming that she had engaged in a flagrant hamburger harangue and caused the "Oprah Crash," costing them millions of dollars. We'll lasso this little Chicago lass, they thought, drag her down to the Texas Panhandle where beef is big business, pick us a dozen tumbleweed-tough locals to serve on the jury, and shut her mouth for good about contaminated beef.

They might as well have walked out and hugged a cactus as

picked on Oprah. In January 1998, she strode into Amarillo as full of smiles and strong attitude as she brings to her daytime show—which, just for good measure, she also brought with her, broadcasting from the Feedlot Capital of the World for the six-week duration of the trial. Beef might be king, but Oprah is queen, and the people of Amarillo flocked to her side. Not only did they scramble to get tickets for her show (even the mayor said that he really hoped he'd get to meet her, and that his wife was going to her show), but more importantly they understood that these bigshot feedlot financiers were talking nonsense when they claimed that her honest statement on television had anything to do with cattle prices: "People that buy and sell cattle are not at home watching Oprah Winfrey," noted a commonsense local standing in line for her show.

After a ton of testimony was brought against her (including that of a former USDA official who was paid more than $25,000 by the feedlot operators to castigate her), the jury needed only seven hours of deliberation to exonerate her completely. After the verdict, juror Pat Gowdy told the press: "We felt that a lot of rights have eroded in this country. Our freedom of speech may be the only one we have left to regain what we have lost."

It was indeed a bright day for our First Amendment rights against the small-mindedness of corporate powers that want to stifle individual freedoms. Unfortunately, the judge did not give the jury a chance to decide on the constitutionality of the Texas food disparagement law, having ruled midway through the trial that it simply did not apply in this case. So the Texas law, and a dozen others like it across the country, is still on the books, still threatening those who oppose the food industry's madness. Next time, you can bet the industry won't pick on someone with deep pockets and a television audience of 20 million viewers.

Still, I have no doubt that these laws will be struck down in due time by the Supreme Court (based on a strict interpretation of the First Amendment's Anticlowning Clause). The laws are not a testament to corporate strength, but to pathetic weakness. They reveal just how touchy and vulnerable today's industrialized, conglomerated, massively centralized, high-tech food producers have become to any public scrutiny of their manufacturing methods.

HOGS

"Old MacDonald had a farm, e-i-e-i-o/And on that farm he had a hog, e-i-e-i-o."

But it's not Old MacDonald delivering your bacon these days. He's been shoved out as more and more hog production is falling under the control of conglomerates with names like OMcD Inc., which control production from semen to supermarket. These computerized meat factories do not have "here a hog, there a hog" running around a barnyard, but more than 100,000 porkers a year jammed together on the concrete floors of hog warehouses that stretch longer than a football field.

Hogs (how shall I put it?) . . . excrete. A lot. *Muy, muy mucho* manure. Pound for pound, they generate nearly twice as much waste as beef cattle do.

This output is not a problem when, say, a thousand farm families in a wide region have a hundred hogs each. The manure of these 100,000 hogs can easily be handled by the families, whose farms are spread across several counties and can absorb the waste. But put all those hogs in one place and you have beaucoup stink, not to mention contamination and health problems. Think of one hundred houses in a large, suburban neighborhood, each one with a family cat. This is fine. Now think of the weird cat lady who keeps one hundred of them in one house. You see the problem.

But confined hogs pose a worse problem because of the volume and poisonous content of their waste. In the crowded building where these animals spend their short lives, they are constantly fed to add quick bulk and send them to slaughter ASAP, with computerized machines steadily and automatically dispensing to each hog a diet heavy in protein, artificial growth stimulants, antibiotics, and other additives. It is a diet that generates millions of gallons of watery waste loaded with nitrogen and phosphorus and laced with copper, zinc, and other heavy metals, along with other chemical additives that poison the soil and water as they accumulate.

Not to worry, say all the OMcD Incs., we drain all that urine, feces, and whatever into giant lagoons right on our own property, and we comply with all the state and federal environmental laws that pertain to our confined-feeding operations.

Sounds swell, except that the states and the feds have *practically zero regulations* that pertain to these huge hog factories, largely because pork has plenty of political muscle that it flexes early and often to prevent lawmakers from doing anything to protect people, even when the community is endangered. For example, folks around Richlands, North Carolina, raised a stink for months about the sickening odor and swarms of flies emanating from Oceanview Hog Farms Inc., a massive facility situated just off the New River, which flows directly into Wilson Bay and on into the Atlantic. But no one in charge did a thing to help—not the county officials, not the governor, not the congressional delegation, not the EPA: "Nothing we can do," they all said, "Goodbye and good luck."

Unfortunately they got bad luck. After a heavy rain in June 1995 Oceanview Inc.'s eight-acre waste lagoon burst open at one corner, and a putrid tide of hog excrement flowed knee-deep across the highway, flooding homes, a churchyard, soybean fields, and everything else in its path, then it poured into the New River. As the bumper sticker says, "Shit Happens." Twenty-five million gallons of it in this case—*more than double the volume of the* Exxon Valdez *oil spill.*

The raw, reddish-brown stinking sewage spread all the way down the New River, from Richlands to Wilson Bay, seventeen miles, killing every fish, plant, and other speck of life that existed in the river.

Oceanview Inc.'s spill is the worst so far, but hardly the only one. In the six months following it, five other major lagoon spills occurred in North Carolina alone, totaling more than 30 million gallons. This does not count hundreds of leaking lagoons or the farm managers who deliberately and illegally dump their hog slurry into the nearest river. In North Carolina in just one summer, 1995, some 10 million fish were killed in the state's rivers, 364,000 acres of bays and marshes were closed to shell fishing, and hundreds of fishing and tourist businesses were shut down—thanks to hog pollution. In addition this nonstop flow of waste constantly threatens people's drinking water with contamination by the parasites, bacteria, viruses, and nitrates it contains.

Water pollution, though, is not the half of it. A plague of flies breeds in these lagoons, and when they swarm there are so many of them that they form living, buzzing clouds, forcing everyone indoors. Another by-product of hogs is acid rain, created by the hydrogen sulfide and other gases emitted in unimaginable volume every second of every day from the hog houses and lagoons.

But the worst is the stench. Oh, *the odor!* The smell spreads for miles around these factories. It is not merely a horrible stink, but a sickening one, a stink that literally sticks to you—your clothes, skin, hair, nostrils—and you cannot wash it away or get away.

Ask Julie Jansen. In 1989 she and her husband, Jeff, bought a beautiful eight-acre place near Olivia, Minnesota, an hour west of Minneapolis—a wonderful spot to raise their six children. And with six kids and eight acres, why not bring in some more children, so she opened a day-care center in her home. Five years later, though, two hog factories were built about a mile away, and her dream place became a nightmare. The whole family began suffering nausea, dizziness, blackouts, chronic headaches, recurring sore throats, nosebleeds, and constant fatigue.

"The kids won't go outside and play when the stink is bad," Julie told *U.S. News & World Report.* "We can't barbeque, hang clothes on the line, sit outside and enjoy the garden. We keep the windows and doors shut, the air conditioner running, but the smell gets in the carpet, the curtains, the furniture. When it gets really bad, we spend the night in a motel. I've had to close my day-care business because nobody wants to bring their children here. We'd like to sell the house and move, but who would buy it?"

Welcome to Hog Hell. You need not travel to Minnesota or North Carolina to experience it, either, because these malodorous factories are headed to your state, too. Already they are in Illinois, Iowa, Kansas, Missouri, Oklahoma, Tennessee, Texas, and Utah, and they are spreading fast as idiotic public officials (who invariably live far away and upwind) provide the special tax breaks and outright subsidies that make hog factories lucrative investments. In recent years foreign syndicates have discovered that they can take advantage of both the subsidies and the lax environmental regulations on these enterprises to come to America and make a killing in pork. Way up at the top of the Texas Panhandle, for example, outside the town of Perryton, an outfit called Texas Farm Inc. got government subsidies to build a massive hog network that explodes the previous concept of "huge," planning to house 2 *million* oinkers in its facility. "Texas

Farm" is not Texan at all, but Japanese. It is a wholly owned arm of the giant Nippon corporation, which is happy to exploit this rural county like a Third World colony, producing pork on the cheap, leaving us with the mess and the health problems, while it exports all the meat back home to Japan's high-dollar market.

TURKEYS

Mark Twain noted, "When we remember we are all mad, the mysteries disappear and life stands explained."

One of nature's creatures that has no doubt whatsoever about the complete madness of us *Homo sapiens* is the turkey, the once-proud forest bird that now finds itself rather grotesquely reconfigured by geneticists to suit the marketing ploys of poultry processors. Years ago, you see, the industry learned that many consumers prefer the white meat of the breast. If white breast meat is in demand, thought the marketers, eyes twitching involuntarily, why not have engineering make us a turkey that has lots of white breast meat, maybe even a *humongous* amount of white breast meat, maybe so much that these turkeys could get jobs at Hooters?

Hence, your Butterball. Today's typical supermarket turkey is twice the weight it was thirty-five years ago, and most of that increase is in breast, which now swells out to near basketball size. This has made the turkey breast lover happy, but, alas, it has destroyed the sex life of turkeys themselves. It is a simple matter of physics—with such breast protrusion, the male literally is unable to reach the female, sexually speaking—and for a while, it looked like the clever breeders had bred themselves right out of business. To deal with this avian sexual dysfunction, the turkey technicians first tried to devise a kind of kinky mechanical solution: They built little saddles to strap on the females so the toms could . . . well, mount up.

This did not work, and it was looking like the BTTI (Big Time Turkey Industry) was at a dead end. Then a couple of California poultry profs perfected the precise practice of *artificial turkey insemination*. Henceforth, turkeys did not have to mate, because we humans could do that for them, so let the mutations multiply!

And so they have. It is not the red, yellowish-brown, and green-feathered turkey of Pilgrim fame that we find plucked, trussed, and frozen in the supermarket today, but a single breed of "white" turkey. Ninety percent of the half-billion turkeys sold in the world each year are derived from only three breeding flocks that are maintained on secretive, highly guarded farms surrounded by chain-link fences. Owning these three flocks, and thus controlling the world market of breeding turkeys, are Merck & Co. (the pharmaceutical giant based in New Jersey), Booker PLC (the British food conglomerate), and British Petroleum (the world's fourth-largest oil company).

All three flocks are of the white breed, assuring such dominant commercial growers as ConAgra Inc. (owner of Butterball) that each and every one of their turkey chicks will uniformly develop those pumped-up breasts. But such genetic uniformity has its ugly side. Not only can these hapless birds not mate, they also are bred to be so heavy and are so disfigured that they can barely walk more than a few feet, and many cannot even stand on their own two drumsticks, so they spend their abbreviated lives mostly squatting in the sawdust of dimly lit turkey houses, jammed wingtip to wingtip with hundreds of their genetically altered siblings. This breed cannot survive on its own, so the birds must pass their entire existence in environmentally controlled buildings, where machinery automatically dispenses a steady ration of feed that is rich in artificial growth stimulants, but often denies them such basic minerals as iron. (To create "a whiter white meat," iron is eliminated from the diet because it imparts a healthy reddishness to turkey flesh and, well, this is not what the marketing department ordered.)

Confined, crowded, and genetically manipulated to meet industry's convenience, these unfortunate creatures are unnaturally vulnerable to infections and disease. Indeed, the world's population of white turkeys is so limited in genetic diversity that an outbreak of some virulent, mutating bacterium is always a threat to devastate the entire breed in a flash. Industry's answer? Antibiotics. And more antibiotics. And when the bacteria adapt to these, get yourself some more powerful antibiotics.

Such farms have become pharmacies and, as one turkey producer notes, today's corporate "pharmer" operates "with syringe in hand," applying dose after dose of drugs, day in and day out. Factory-produced animals now receive an average of about thirty times more antibiotics than people do. But since residues of these drugs can end up in the meat, you can end up as doctored as the turkey, getting an unexpected dose of the bird's antibiotic mix with every bite of Butterball you take.

What you bite into, by the way, is not the savory richness that turkeys naturally have, but the synthetic blandness of mass production. Marian Burros, the cookbook author and food columnist of the *New York Times*, conducted a blind taste test of nine turkeys just before Thanksgiving last year. The favorite by far, she reported, was an organically produced turkey—rich and meaty. Staggering in for last place was ConAgra's Butterball—"Tastes like chemical stew," was the judgment of the tasting panel. No wonder. In addition to its bizarre breeding and factory raising, the poor bird is injected after slaughter with vegetable oil, water, salt, emulsifiers (mono- and diglycerides), sodium phosphate, annoto color, and artificial flavor. No butter, in case you are wondering; more like a partially hydrogenated Oilball.

Add to this the stuffing problem. Cooks are now warned by the Department of Agriculture not to stuff their Butterballs. Because the birds are so breast heavy, having been genetically loaded with the quicker-cooking white meat, the roasting process is not always long enough to heat the stuffing to a temperature adequate to kill the bacteria—giving an unhappy meaning to the term "chef's surprise."

A more appropriate use of these torqued-up turkeys might be the one enjoyed by the fun-loving gang down at DT's Saloon, just outside Cleveland. They bowl with them. A length of floor adjacent to the bar has been marked off as a bowling lane, wooden pins are set up at one end, and DT's happy imbibers go at it, slinging ten- to twenty-pound frozen butter-"balls," still in their plastic wrappers, down the lane.

MAD COWS

Years ago General Motors decided to promote its Chevy Nova in Puerto Rico, so it developed an enthusiastic advertising effort to persuade the people of this Spanish-speaking island to rush out and buy the fabulous new car: La Nova! Problem is, in Spanish, *no va* means "It does not go."

I always wondered what happened to the creative team that came up with GM's "It does not go" promotion. I assumed they had been bound and gagged, placed in an old Nova, and sent to the scrap-metal yard, where giant crushers compacted them and the car down to approximately the size of a pack of Juicy Fruit. But now it occurs to me that maybe they got away with it after all, changed their identities, and became federal bureaucrats in charge of our nation's food-safety regulations. How else can we explain the fact that whenever America confronts a crisis of contamination in our food supply, the response from those running our regulatory mechanism is always "No Va"?

Consider the distressing case of cow cannibalism. The story begins in the fall of 1986 in Kent County, England, where veterinarians first began detecting cattle infected with BSE, or bovine spongiform encephalopathy—which quickly became known popularly as mad cow disease. BSE is an infectious, incurable, neurodegenerative affliction that gradually rots a cow's brain, literally causing the animal to lose its mind—then it dies.

But what lifts this beyond the genre of "sad cow story" is that BSE-infected cattle are sent to slaughter long before they show any outward symptom of the disease. And since existing methods of meat inspection cannot detect the malformed proteins that carry BSE, the infected beef is cut, chopped, ground, packaged, and shipped to market with a nice "Grade A Approved" sticker applied by the government's unknowing inspector. The nation's unsuspecting consumers then swallow the government's assurances and, possibly, a deadly dollop of odorless and tasteless BSE.

"Oh pish, posh, and poppycock," sputtered England's Keepers of the Official Order (KOO) when critics there sounded consumer alarms back in the '80s. "Mad cow disease is just that," they blustered, "a cow disease, not transferable to people at all, and certainly not to properly bred Englishmen. Stiff upper lip and all that, old boy."

In 1986 there were seven BSE cases reported in England. The next year, 413; the next, 2,185; the next, 7,136; the next, 14,180; the next, 25,025 . . . and upward and upward—at least 160,000 British cattle known to have BSE by 1996. Still the KOO kept insisting: "No problem," denouncing as alarmist left-wing balderdash the desperate warnings by a courageous few scientific critics that a BSE-epidemic was in the making and that humans would die. In the time-honored tradition of officialdom everywhere, British agriculture minister John Gummer even resorted to going on the telly in 1990 with his adorable four-year-old daughter to chomp hamburgers together. "It's delicious," Minister Gummer officially pronounced between munches, and a nation of viewers silently responded: "You're a blithering idiot, John."

He is an idiot. The human equivalent of BSE is CJD—Creutzfeldt-Jakob Disease—a 100 percent infectious, brain-wasting, fatal dementia that produces a truly horrific death. Supposedly a rare disease, CJD cases suddenly began to rise in England in the early 1990s, doubling in the next four years. "No connection," the KOO sniffed.

Then came the day of reckoning, March 20, 1996. The minister of health, who had previously been adamant that mad cow disease posed no danger whatsoever to humans, appeared ashen-faced before the House of Commons to report that well, er, ahem, ah . . . *oops!* The minister confirmed that ten new cases of a variant CJD had cropped up, including a cluster of them in Kent County, and that mad cow disease, having been transferred to the victims through hamburgers and other beef products, appeared to be the cause. In Glasgow a professor of neurology overseeing a young girl's affliction said flatly: "She has BSE-pattern CJD and picked it up through hamburgers." He added, "The potential for developing this disease must reside in millions of the population, because I am sure there are many people who have eaten infected meat."

The scale of the epidemic is as yet uncertain, because if you eat a bad hamburger and are infected with CJD, you will not know it for years. The disease can incubate in your body for between ten and fifty years with no outward sign of trouble—then, whammo, your brain is riddled with holes, your nervous system comes apart, you become demented, and you die. Painfully and grotesquely. There is no way of knowing how many Brits have already consumed contaminated meat, but just as the number of BSE cases has risen dramatically since 1986, so are the number of CJD cases expected to rise from 1996 forward. Dr. Richard Lacey, one of England's top microbiologists and the country's most insistent scientific critic of the KOO's fatal policy of denial on BSE, correctly predicted years ago that the first human outbreak of mad cow disease would occur in 1996, and he now forecasts that more than

5,000 British people a year (and perhaps as many as a half-million a year) will be dying of it in the next century. Dr. Lacey declared, "This is one of the most disgraceful episodes in this country's history. The government has been deliberately risking the health of the population for a decade."

How now, mad cow? How did BSE become epidemic and migrate from bovines to us bipeds?

Through cannibalism. Cows are natural herbivores—contented, cud-chewing vegetarians that gladly would pass their days grazing a field of grass, or sustaining themselves with nothing more complex than a bale of hay. But modern agribusiness has no patience with a cow's natural rhythms, insisting instead that tens of thousands of them be jammed together in feedlots and massive dairy factories, where they are force-fed a diet heavy in fat and protein "supplements" to goose up their milk and meat production.

Through these supplements, industry has turned natural herbivores into carnivores, for the supplements are made from ground up animal parts, including brains, spinal cords, spleens, tonsils, tails, odd bits of flesh, grease, nerve tissue, blood, bones, and other offal left over from the slaughtering of cattle. In other words, industry has been feeding cows to cows. There was a delicious twist during the Oprah Winfrey mad cow trial when a witness for the feedlot operators attempted to deny that "cows are eating other cows." He lamely tried to tell the jury that eating such cow parts as rendered bone and meat is not the same as a cow actually eating a cow. But in her subsequent testimony on the stand, Oprah reduced his argument to BS, saying: "I have to tell ya, to me, whether it's ground hamburger, whether it's powdered, whether it's cow tea . . . it's still eating ground cow."

The remains of sheep, goats, and other animals susceptible to their own forms of BSE also have been put into the rendering vats, melted down, and processed into protein pellets sold as cattle feed, as well as feed for hogs, chickens, household pets, and other animals. What industry and government have known, but not shared with the rest of us, is that the rendering process does not kill BSE. Indeed, little does kill it—BSE-infected matter drawn from its victims has remained infectious even though it was bombarded with radiation, soaked in formaldehyde, or baked at 700 degrees. This virulent disease is delivered directly to live animals through the cannibalistic protein supplements they are fed. As little as a teaspoon of feed rendered from diseased cattle can transmit BSE to others.

One cow leads to another, and before you know it you have a BSE epidemic, a CJD outbreak, an angry and alarmed public, a devastated beef and dairy industry, and a disgraced government. Britain finally banned the industry practice of feeding rendered cow parts to cows in 1989, but animals born before then are suspect, so the government is now embarked on a campaign of mass incineration, a nasty, five-year project of bovinicide that will cost taxpayers more than $3 billion, and cost some 5 million innocent cattle their lives. All of this has occurred because modern agriculture's profiteers tried to beef up their beeves and their profits with a quick dose of cannibalism.

One bit of irony to come from industry's attempt to convert herbivores to carnivores is that the resultant disease and public health scare converted masses of British carnivores to herbivores. For months beef went untouched in supermarkets, meat disappeared from school cafeterias, fast-food burger joints advertised "Foreign Beef," tofu sales were suddenly hot—and the public was not bothered by any more appearances on the telly by agriculture ministers munching meat and mumbling inanities.

But this is not merely a saga of mad cows and Englishmen. Across the big drink, American cattle feeders, dairies, and factory farms have been practicing cow cannibalism on a far more reckless scale than the Brits, and our public health protectors have been in even deeper denial about the high price we consumers are paying for agricultural industrialization (including the likelihood that an American version of BSE already is loose in our herds and a possibility that related-CJD is loose in our population).

As independent ranchers have been pushed aside by corporate feedlot managers in the cattle business, and as "old Bossie" has become nothing but a number in a dairy industry run by faraway corporate bosses, the use of slaughterhouse waste to feed America's beef cattle and dairy cows rose precipitously. Currently, of every cow that is hauled to slaughter, about 14 percent of its weight ends up as protein supplements fed back to other cows. Sheep offal goes into the mix, too, even though American sheep are known to carry "scrapie," the same form of spongiform encephalopathy that figured prominently in England's BSE epidemic.

Then it gets ugly. In addition to using slaughter waste to make feed supplements, the other stuff that goes into the renderers' vats would gag the witches of *Macbeth*. Like roadkill. Yep, that plastered possum, that dead armadillo in the middle of the road, that deer caught in the headlights—all fodder for cattle feed. Dead cats and dogs find their way into the vats, too. For example, the city of Los Angeles sends 200 tons of euthanized cats and dogs to an L.A. company called West Coast Rendering every month. Also tossed in is a category of farm animals delicately referred to in the industry as "4Ds." These include cattle, hogs, sheep, and others that are "Dead, Dying, Diseased, or Disabled." Got one of these on your farm? Call the renderer—the American Feed Industry Association estimates that nearly 400,000 tons of 4Ds are used each year to make animal feed.

Does feeding this mad brew of animal flesh and body fluids back to cows pose any health hazards in our country? Is a pig's butt pork? As the late Dr. Richard Marsh, professor of animal health and biomedical sciences at the University of Wisconsin, so plainly put it: "We have all the risk factors. Just because [mad cow disease] hasn't been diagnosed doesn't mean we're not somewhere in the cycle. It's obvious. I mean, we have [the disease] in our sheep and we feed sheep to our cattle. What's the argument about?"

Elvis answered that one for us long ago: "Money, honey." Beef is a $60-billion-a-year industry here, milk does $29 billion, the animal feed industry rings up $20 billion and the rendering companies haul in 3 billion bucks a year selling their particles of yuck to the animal feeders and to the pet food companies. Unlike consumers, these people have real power at the USDA, FDA, APHIS, CVM, EPA, USPHS, and all the other nonsensical scrambles of Scrabble tiles that pose as the public's regulatory protectors against industrial abuses.

It was no surprise, then, that American officialdom rose as one to defend the purity of our beef and dairy industries when that nasty bit of BSE business hit the English. As if to prove that the Brits have no patent on Goofiness in High Places, the sitting Texas agriculture commissioner even darted out on April 4, 1996 (barely missing April Fools' Day, darn it) to do his best imitation of British ag minister Gummer's hamburger-munching routine. At a hurriedly assembled press conference, the Texas commish chawed on a big chunk of smoked brisket for the TV cameras and, through greasy lips, pronounced it "delicious" and "safe," thereby convincing most viewers that the poor fellow undoubtedly was suffering from a well-advanced case of mad something.

Not that U.S. officials were caught unaware by the BSE flare-up. In 1989, when the British banned cow cannibalism, the ever-vigilant U.S. Department of Agriculture took a bold bureaucratic stand, too: It appointed a committee. Verrrrry quietly, so as (shhhhh) not to disturb the consuming public. Dubbed the "BSE Consultants Group," its members are drawn from the National Cattlemen's Association, National Milk Producers Federation, American Sheep Industry Association, and National Renderers Association. Notice there are no consumer groups listed. To this day no consumer advocacy organization has been asked to be part of the BSE discussions.

However, thanks to a Freedom of Information action by the Pure Food Campaign, a consumer watchdog group, we can get a glimpse of USDA's deeply felt concern on this critical issue—concern for protecting the industry, that is. In a 1991 USDA memo that the group unearthed, the ag agency acknowledges the possibility that the disease "is present in the U.S. cattle population," then notes that if government was to "prohibit the feeding of sheep and cattle-origin protein products to all ruminants" it would minimize the risk of spreading the disease. But, the memo concludes, "The disadvantage is that the cost to the livestock and rendering industries would be substantial." True to form, USDA imposed no ban.

Far from it. Instead of protecting the public health, the agency drew up a contingency PR plan to protect the industries' image if and when mad cow disease breaks out in our country. In a second memo with the unusually candid title of "1991 Bovine Spongiform Encephalopathy Public Relations," our so-called public protectors warn that "agriculture is vulnerable to media scrutiny [on] the practice of feeding rendered ruminant products to ruminants and [on] the risk to human health." According to the memo, USDA staff had analyzed British press clips to determine what "strategic errors" they had made and to help industry here "avoid public relations problems such as have occurred in the UK." The memo chastised British officials for even trying to keep track of CJD cases, for example, because

"This appeared to legitimize concern about a link between BSE and human health." The memo's finest flourish of PR flackery, though, is its pitiable effort at language control: "The term 'mad cow disease' has been detrimental. We should emphasize the need to use the term 'bovine spongiform encephalopathy.'"

Amateurish as this whitewash seems, the conventional media has swallowed it whole. "Federal officials are confident that the mad cow disease found in Britain has not occurred in the United States and that existing policies are adequate to protect the beef supply," asserted the *New York Times* lead paragraph, unquestioningly, just one week after British officials had been forced to admit that the exact same assurances they had been giving their public for years were untrue.

To know the truth, one had to be reading articles in the alternative press, especially the work of Joel Bleifuss in *In These Times* and, more recently, of John Stauber and Sheldon Rampton in their scrappy newsletter, *PR Watch*. There you would have learned about the contrarian findings of Dr. Marsh. This highly regarded research scientist died in March 1997, but not before his uncompromising work in the labs and in the field caught the industry and government in their self-serving lies about mad cow disease, and not before his sharp, insistent voice of scientific protest had punctured every PR balloon they had floated about this deadly issue:

Industry/Government: BSE does *not* exist in the United States, and we maintain a thorough program of testing cattle herds to make certain that an outbreak never occurs.

Dr. Marsh: The government's testing program has only been in existence for five years. In that time there have been as many as 500 million cattle in the United States. A *total* of only 6,000 of them have been tested for this disease. Even then, officials have asked the wrong diagnostic questions, looking only for brain damage identical to British BSE cases, ignoring the likelihood that the American version presents different symptoms.

Industry/Government: We have even tried to replicate the disease by exposing U.S. cattle to a version of it, but they simply did not go mad à la the bovines of Britannia.

Dr. Marsh: The American strain of BSE shows different symptoms but is just as deadly—instead of madness, the diseased U.S. cows simply fall down and die. A team of scientists, including me, have proven this in USDA-sponsored tests in Iowa and Texas, injecting cattle with brain matter from infected sheep. Sure enough, the cows fell down and died. BSE-infected cows, then, could be part of the 100,000 cattle a year that die from a phenomenon known as "Downer Cow Syndrome."

Industry/Government: Well, maybe so, but no Americans have died from BSE.

Dr. Marsh: How do you know? The Centers for Disease Control did not even bother tracking cases of Creutzfeld-Jacob Disease in our country until April 1996, after the outbreak in England. Even now, the Center only checks on CJD cases in five states, and the "checking" they do in these is woefully inadequate.

Choose your own expert, but I go with Dr. Marsh. Indeed, more and more medical studies of dementia suggest that some of the 4 million Americans diagnosed with Alzheimer's disease are actually suffering with CJD, which has remarkably similar symptoms. It is now estimated that there could be as many as 40,000 cases of CJD in our country today—not the 250 or so cases claimed by officialdom.

Still, Washington denies and dawdles. Despite the urgent pushing by Dr. Marsh and others for government to get serious about protecting the public health, the regulators and industry have been so tight you couldn't get them apart with a crowbar. In 1992, for example, Marsh called on the Consultants Group to expand its research focus on BSE to include Downer Cow Syndrome, but the

Group refused, saying that it had "agreed that the current efforts are on target for the needs of the livestock and rendering industries . . . and that changes in the research direction are not appropriate at this time." Even after the British epidemic, USDA officials stayed hunkered down like cows in a hailstorm, refusing to take the obvious step of banning cow cannibalism, as first suggested five years earlier. A report from the April 1996 Consultants Group meeting says it all: "The cost of banning ruminant feed now versus later depends on risk factors which should be determined by science, not emotion."

Whose science? The science of the corporate interests dominating the Consultants Group, of course. Anyone else's science is, by definition, "emotion." Based on the industry's "science," the BSE panel decided that, in response to the manure pile of evidence built up by the British disaster, maybe a nice *voluntary* ban on cow cannibalism in this country would be enough to convince us emotional consumers that everything is just hunky dory. So, in the spring of '96, the Clinton Administration's official response to the real-life threat of BSE was to sanction an industry promise that it would police itself.

One wonders, are they stupid, or do they just think we're stupid? Consumer protection through volunteerism is no more effective than a Jell-O doorstop—even the top scientist for the Renderers Association conceded to the *Boulder Weekly* a few weeks into the highly publicized "ban" that he could not think of a single company that was complying with it. "Voluntary bans don't work very well in a free-market enterprise," he deadpanned.

There is a band in Austin called Two Nice Girls. It is a four-girl band. That's a fun joke for a band, but the joke is on us when it comes to the halfway mentality of those in charge of protecting public health. Actually, the craven compromises of the Clinton Administration are not even half-good, much less nice, for they so often compromise the well-being of our families for nothing more noble than an extra nickel's profit to a handful of special interests. Especially insulting is that these compromises, as in the case of cow cannibalism, are for things the corporate interests do not even need. In July 1993, for example, the *Food Chemical News* reported that a spokesman for the National Cattleman's Association admitted to the

BSE Consultants Group that "his industry could find economically feasible alternatives to the 15 percent of feed now supplied by animal protein." So why didn't they do so? Because, reported the *News*, "the association does not want to set a precedent of being ruled by 'activists.'" Great. We get exposed to deadly disease just to satisfy the political agenda that an industry group is running against consumer activists.

Four years after that memo the games are still being played, even though something happened late last year that finally forced our "protectors" in Washington at least to pretend they are doing something. On October 24 British scientists reported that they had found a protein in the brains of humans who had died from CJD that was almost identical to the mad cow protein, providing alarming scientific evidence that the cow disease appears to have jumped to humans, directly challenging the cascade of claims by industry and government that this was not possible. At last, then, in January 1997, the Food and Drug Administration took the long-overdue step of proposing a government ban on cow cannibalism.

Do not sigh with relief just yet, though. Note the verb: proposing. This does not mean *im*posing or otherwise actually doing something. On June 5, five months after the *proposed* ban, the FDA quietly issued its actual rules on feeding rendered mammal parts to the mammals we humans eat. The "ban" turned out to be greasier than a rendered warthog, thanks to the slick lobbying of the meat and rendering industries and thanks to the Clinton Administration's infamous obsequiousness to corporate interests:

> First, the FDA's ban on feeding cow tissue to other cows is only a *partial* ban. It exempts the blood and gelatin (derived from hides) of cows, even though both are suspected by World Health Organization scientists of carrying the deadly proteins that create the disease.
>
> Second, the "ban" also specifically exempts pigs, meaning their rendered remains can be fed to cows, even though there is increasing evidence that America's pork herds could be infected with their own form of BSE—a sort of mad pig disease.

> Third, while the rule says no rendered cows or other ruminant animals can be fed to cows, protein pellets made from these ruminants—even those *known* to have BSE—still can be made and fed to pigs, chickens, fish, pets, and other animals, thus providing a clear path for BSE to enter the food chain.
>
> Fourth, there is no way to enforce the ban against feeding these cow pellets to cows. Anyone can buy the pellets. Yes, they are labeled "Do not feed to cattle," but that is no more of a deterrent than signs at the seashore saying "Do not feed the seagulls."

As Dr. Michael Hansen, a research associate with Consumers Union, puts it, "Given the difficulties of reversing a [BSE-style] epidemic, it seems to us the best course is to take all necessary steps to keep one from getting started." Thank you. Why don't we put *him* in charge of the FDA? The only sure way to prevent a menagerie of mad meat diseases from sweeping our country is to place a flat ban on the rendering of any mammal into feed for any animal used for human food or even for pet food (that is unless you plan to heat the spoons you use to dish out Fluffy's and Fido's chow to 750° Fahrenheit). In other words, stop the cannibalism, and stop it where it can be enforced—at the renderers' vats. Anything less is playing Russian roulette with the health of our families.

Some jokester once claimed that stupidity is not terminal because it produces antibodies, but anyone who takes a look at what our industrialized civilization is producing in the way of hogs, turkeys, and mad cows can get a glimpse of terminal stupidity in full roar. The industrialists and money managers of modern agribusiness have fabricated a factory-food system that messes with Mother Nature something awful. It would be one thing if their madness produced better or cheaper food for us, but it does not. Indeed, in terms of our health and our environment, your family and mine are only beginning to reap the bitter fruit of their methods. No longer is dinner a safe bet, and no longer can public authorities ignore the bizarre and virulent parasites, the aggressive carcinogens, the estrogen-mimicking chemicals, and other unwelcome additions to our food supply. These are hardly inadvertent or uncommon occurrences—

rather, they are inevitable by-products that industry has built into its production system.

It would be comical, were it not so pathetically tragic, to see our public officials keep trotting out with stern advisories for us consumers to cook meat thoroughly as a "precaution" against *E. coli* 0157:H7 or to scrub our strawberries with warm water and soap in the vain hope of cleansing them of pesticide residues and exotic bacteria. What they have done to dinner is beyond the remedies of home ec instructors. It is the *industrial processes themselves* that must be confronted—and changed.

As Kin Hubbard, a turn-of-the-century wit, put it: "If the government was as afraid of disturbing the consumer as it is of disturbing business, this would be some democracy."

UNCLE BEN

Apparently my Uncle Ben Fletcher was a fool.

Actually he was my great-uncle, already an old man by about 1950, when I first recall my brother and me visiting him and Aunt Emma on their little truck farm outside Weatherford, Texas, due west of Fort Worth. At the time we had no idea Uncle Ben was a fool, and we probably would have jumped anyone who said he was. In our boyish innocence, Jerry and I thought Uncle Ben was clever as could be—he was brimming with pranksterish fun, always had a story to tell, could pitch washers like nobody's fool, never dressed in anything tighter than bib overalls, raised everything from chickens to watermelons (and a family, too) on that thirty-acre hardscrabble plot, was blessed with a quick and exuberant laugh, and plainly enjoyed life.

By the standards of modern agriculture, though, he was a fool of a farmer, failing to make use of the full arsenal of petroleum-based inputs and chemical products today's agribusiness employs to maximize profits. Still, this failure and fool did manage to make a crop, as did his neighbors, supplying the whole area with a terrific variety of succulent fruits, flavorful meats, fresh vegetables practically bursting with nutrition and taste, rich milk and cheeses, hearty grains, and . . . well, a cornucopia. God, the food was good!

And no one who bought it had to wonder or worry if any of it contained such contaminants as pesticide residues and genetically engineered hormones, because Uncle Ben and the others raised this abundance without a trace of chemicals. Imagine. The fools simply did not know any better.

Farmers have been taught a lot since then, of course. In only a generation, Uncle Ben's plow was replaced by the chemical sprayer as the apt symbol of agriculture. Thanks to the introduction of synthetic fertilizers, insecticides, herbicides, and fungicides, instead of a hundred Uncle Bens nurturing crops on thirty acres each, we now have one farm operator managing 3,000 acres . . . and trying to manage the big bank debt incurred to underwrite such high-tech production. This is a result that agricultural economists (who have the collective vision of dung beetles) cite in their learned journals, lectures, and chemical company consultancy reports as the very model of agricultural "efficiency."

Theirs is an efficiency that conveniently factors out what is known in the jargon of economists as "externalities" (Literal Translation: big, ugly numbers that would cause our equations and conclusions not to add up, therefore we shut our eyes tight as can be and pretend they are not there). But in your and my real-life environment, unlike in their EconomicsWorld, externalities are hard to ignore. For example, the sheer volume of chemicals being used in their "model of efficiency" crops up in the most unpleasant ways, costing far more than our good earth, farmers, and progeny can stand to pay. Since the mid-sixties, pesticide use on farms has doubled, with nearly a billion pounds of active ingredients now being applied each year. Another 4 billion pounds of "inert" chemicals are added to the pesticide mix, too, including known cancer causers and other toxics, though we are not told which ones are in which blend (not only does federal law allow pesticide makers to keep their list of "inert" ingredients secret, but also a $10,000 fine can be levied against any federal employee who spills the beans and discloses what killer chemicals are used as "inerts"). This total pesticide dosage of 5 billion pounds a year on farms is *twenty pounds* for every man, woman, and child in America. Good grief, we consume only about five pounds of butter per capita, and eight pounds of coffee (and the health patrol gives us a hell of a tongue-lashing for this excess), yet there goes agribusiness merrily peppering our dinner with twenty pounds of poisons apiece.

We pay for this—in taxes, in illness, in health-care costs, and in suffering. For example, farm pesticides run off into our drinking water—nearly a hundred different pesticides are now found in the groundwater of forty states, contaminating the essential liquid of life for 100 million of us. Our taxes have to be increased to try to filter out some of these toxics before they reach our taps, then we have to pay even more dearly for the crushing health costs—monetary and otherwise—of the poisons that are not filtered out.

Frontier communities had a quick and certain remedy for anyone who poisoned the town's well: They hanged the son-of-a-bitch. Today, though, when the ag economists draw up their efficiency equations, well poisoning is not even marked down as a cost charged to the poisoners—instead, it's dismissed as an "externality." Did people get breast cancer? Did the pesticides run off into the bay and shut down the fishing industry? Was a farm worker's baby born with birth defects? Hey, pal, stuff happens, life ain't fair, not our fault, get out of the way of progress . . . and if you're so prissy about poisons, maybe you oughta start boiling your water.

Of course, boiling water to eliminate poisons would be as ineffective as the poisons are proving to be against the pests. It turns out that bugs, weeds, and other agricultural pests are amazingly adaptable—some 500 species of insects have already developed genetic resistance to pesticides, as have 150 plant diseases, 133 kinds of weeds, and 70 species of fungus.

Yet, like General George A. Custer—who on his last morning on earth shouted encouragingly to his troops: "On to Little Big Horn for glory. We've caught them napping!"—the industry urges farmers to charge dead-ahead, to apply ever more poisons and ever more poisonous poisons. Ever more expensive ones, too, now costing farmers $8 billion a year, not counting the cost of spreading them. While the pests are adapting and surviving this chemical onslaught, farmers are not—ironically, thousands are killed off financially each year by the escalating cost of the very pesticides that were supposed to make them so efficient. Worse, thousands of farmers, farm workers, and rural residents die prematurely each year, victims of their prolonged exposure to agriculture's "chemical revolution."

Old Texas Saying Number One: "If you find you've dug yourself into a hole, the very first thing to do is to quit digging." The good news is that farmers everywhere are eager to quit digging, but like people hooked on drugs, they find themselves caught up in a *culture* of chemical use, with everyone from the government to the banker pushing more chemicals on them.

Old Texas Saying Number Two: "Where there's a will, there are at least a thousand won'ts." Try to change and you are up against the whole ag system. Start with the extensive rural network of traveling agents for Monsanto, Dow, and the rest who constantly make their rounds, pushing poisons from county to county and from farm to farm. Go to the public agencies—the county extension office, the federal farm research station, or the state ag college—for advice or help, and you'll find them singing hosannas for whatever is industry's pesticide du jour. Ask about organic production and they treat you like you walked in wearing a pink tutu. Politicians are no help, because practically all of them take money from the pesticide peddlers, so they are always Johnny-on-the-spot to defend even the industry's worst excesses and insist that chemicals are nothing less than manna from heaven. Then there's the Farm Bureau, a blowhard agribusiness bureaucracy that falsely fronts as the "national representative of farmers" (most of its members live in big cities). The bureau not only preaches pesticide use relentlessly, but is in the pesticide business itself. Or go to the real power in a rural community—the bank—and you'll get a hard and quick lesson about the industry's grip on agriculture. When farmers seek crop loans from banks and federal lenders, they are handed applications that include a promise that they will follow a detailed schedule of applying chemicals to their land. No poisons, no loan. No loan, no farm. Gotcha.

Still, despite the pervasive power of the chemical industry, despite a nay-saying, foot-dragging, brain-dead agricultural establishment, change is coming—inexorably and with accelerating speed. Agriculture is and will continue shifting to low-chemical use and eventually to organic production because of two irresistible forces: you and farmers.

First, the collective "you"—consumers, the public, the market. All together now: "*We don't want your damn poisons in our food, in our water, in our babies!*" It is not merely a few sprout-eating ex-hippies who feel this way, but the mainstream, the center, the majority—even Republicans, for God's sake. Pollsters confirm that nine out of ten of us are demanding what the trend-trackers have dubbed the Clean-Food Diet, which is described in the *New York Times* as "foods free of artificial preservatives, coloring, irradiation, synthetic pesticides, fungicides, ripening agents, fumigants, drug residues, and growth hormones," as well as foods that are "processed, packaged, transported, and stored to retain maximum nutritional value."

By far the biggest surge in food marketing is toward clean food and natural products: Organic product sales doubled from 1989 to 1994 and now are rising by more than 20 percent a year; overall, the sales of natural products topped $11 billion in 1996; the number of natural-foods supermarkets went from 195 in 1991 to 858 in 1995, and most conventional supermarket chains are now compelled by consumer demand to carry lines of organic foods; organic cotton clothing and linens are becoming all the rage, mass-marketed not only through upscale, enviro-trendy catalogues, but also at Kmart and JCPenney; and if you think the market has nowhere else to go, let me be the first to inform you that a Texas company wants to pamper you with Organic Cotton Toilet Paper. Who says we do not live in exciting times?

"Hrumph," grump the hidebound heads of ag-biz, "organic farming is nothing but a bunch of frou-frous on farmettes, refugees from Haight-Ashbury trying to get you gullible consumers to buy a basket of spotted, bug-infested tomatoes they brought to town in their graffiti-flecked '65 VW bus. Turn farming over to organic producers and half the world will starve, because these dips cannot—repeat *not*—produce the massive yields that our Miracle of Modern Agriculture turns out every day, God bless America and the Chemical Manufacturers Association, end of discussion."

Of course, their adamancy is rooted in the sterility of their argument. It is an argument they are doomed to lose because of the second irresistible force propelling us toward an organic future: farmers. As happens to power establishments everywhere, agriculture's long ago began sniffing its own BS and believing it to be perfume, so it has lost touch with terra firma and those who farm it. To get back in touch, and to see where agriculture is headed, they could do no better than to visit Jim Crawford—one of America's *real* organic farmers.

He doesn't hail from Berkeley, as would befit the establishment's stereotypical profile of an organic producer. Crawford comes from just outside Muleshoe, Texas. As the name suggests, this Panhandle town (population 4,842) is in unabashed, unrelenting farm country, with surrounding towns branded with such names as Hereford, Earth, Spade, Bovina, and Lariat. It is a no-nonsense place where folks are not much given to trendy frou-frou, and I can attest that Jim Crawford is a true son of the place, with more John Wayne in him than John Lennon. A fourth-generation Panhandle farmer, he graduated from Muleshoe High, got an ag degree at Texas Tech just sixty miles down U.S. 84, then returned to Muleshoe to apply what he was taught in college, which was to apply layer after layer after layer of synthetic fertilizers and pesticides to his 900 acres of corn and cotton. He did this for seventeen years, until he concluded in the late eighties that he had had all the conventional wisdom about "how to farm" that he could stand: Jim was on the brink of broke.

With his spreadsheets on the kitchen table and his red pencil in hand, he saw that his costs were only going to keep going up and the price of his crops was not, so something had to give, or he had to give it up. Right then and there is when Crawford became an organic farmer—not through some Earth Day epiphany, but through a hard-eyed examination of the bottom line.

It dawned on him that generations of folks before him had farmed successfully without all those chemicals he was using, so maybe they knew something worth knowing. He began to talk to old-timers, check out books, make calls to what the experts dismiss as "oddball" groups, and generally to study the possibilities—all of which led him to discover that the solution had been right under his nose all the time: manure.

With more cattle than people, the Muleshoe area has beaucoup manure, and Jim's self-education project was teaching him that plowing composted manure or other organic matter into his fields literally would restore life to soil that had been killed with repeated poisoning. The manure begins putting back microbes, minerals, and other natural elements, while also improving the soil's "tilth" so it retains moisture and holds plants more firmly (no small task on these high plains, where the wind blows so constantly that you are advised not to go chasing after your cowboy hat if it blows off, just turn around and a better one will be coming along behind you).

Organic farming is not simply a matter of doing away with chemicals; indeed, farming organically is more complex and difficult than farming chemically, because it requires an understanding of how soil, plants, and pests interact, and it puts a priority on strengthening the soil and nurturing it. Far easier just to nuke the pests, plants, and soil with some 2,4-D and let the devil take the hindmost. At its core, organic farming recognizes that agriculture is the art and science of *cooperating* with nature, rather than always trying to overwhelm it.

Crawford started his shift from synthetics to organics on his corn crop. The first year, just as the snickering nay-sayers of ag-biz had predicted, he got only half the yield he used to get. But the next year, as his manure applications began to work their natural wonders, his yield was nearly back to what it had been, plus he was not spending money to buy chemicals and spread them. Less snickering. By the third year his soil was getting plenty stout and his corn yield was one-and-a-half times better than what his neighbors were getting the chemical way. End of snickering. Plus his corn crop used only half the water others required, and his premium-quality corn earned him a premium price. The snickering was replaced by area farmers stopping by, digging their boot heels into his plowed field, and saying: "Manure, huh?"

Uncle Ben was no fool . . . just ahead of his time.

TIME MAGAZINE IS KILLING US

In McAllen, Texas, way down at the southern tip of the state, you will find the unofficial monument to the paternalism, hubris, and greed of such old Anglo families as the Bentsens, Shivers, and Neuhauses. For decades, up until the mid-seventies, these feudalistic *patrons* pretty much ruled the economy, politics, and teeming Mexican-American majority throughout the Rio Grande Valley.

There is no brass plaque on their monument—it is simply a nondescript twenty-story bank tower, distinguished solely by the fact that it thrusts so phallically out of the region's flat, coastal plain and is visible for miles. This macho edifice was built in the late seventies by Doc Neuhaus, reigning patriarch of his family's fiefdom. He put it up just at the time the valley's old-family structure was coming down—indeed, old Doc himself went to his eternal reward shortly after his "skyscraper" was completed. To this day it stands out in the valley, looming above all beneath it. But instead of standing as a lasting testament to his stature, the structure is mockingly referred to in the now-ascendant Mexican-American community as "Doc Neuhaus's last erection."

Our civilization's last erection may be in sight, too, if we do not finally confront the paternalism, hubris, and greed of the chemical industry, which has had its way with us for way too long. Confronting this poison-spewing Goliath is not merely a matter of laws and regulations, but of changing the way we think about chemicals, challenging the corporate orthodoxy that insists they are essential to modern living.

Excuse a momentary lapse into techno-nerdiness, but let me say this to you: "organochlorines." This is not a term you hear on the nightly news or in political campaigns, yet it embodies a reality with far more destructive impact on your life than, say, the "recidivism rate" of criminals, just to name one technical phrase the system chooses to teach us.

You are not likely to know the organochlorine family by name, but they know you and have an up-close-and-personal presence in your daily life. To give you a quick marker for what we are dealing with, organochlorines were first produced on a large scale in World

War II to make poison gas. Since then some 11,000 different offspring of this chlorine family of chemicals have been created for industry to use in making everything from paint to plastics, pesticides to paper. All 11,000 are toxic: DDT, so deadly that it is now banned in our country, though it remains extensively in our environment, is an organochlorine; so is dioxin, the killer by-product of everything from Agent Orange to hazardous waste incinerators; so are PCBs and vinyl chloride—the basic ingredients of plastic, both known to cause cancer directly; so is Atrazine, one of the most widely used herbicides in North America, a known causer of breast cancer and a major contaminator of our lakes, rivers, and groundwater; so are CFCs, the ozone-destroying chlorofluorocarbons; so is chlordane, another cancer producer widely spread to exterminate termites in homes.

The chemical industry literally pours a toxic stew of these industrial chlorines into our environment, producing about *40 million tons* of them worldwide each year—so much of so many that no matter where you live, you are eating, breathing, and drinking them every day, and your body is contaminated with them. They have been delivered through air streams, fish, ocean currents, migratory birds, and other means of travel into ecosystems absolutely everywhere on our globe—even polar bears in the farthest reaches of the frozen Arctic are now contaminated with them.

The human health impacts of these nasties are only beginning to come home to roost. Despite the fact that this family of chemicals has massively and suddenly permeated our lives and invaded our bodies, government has rarely even studied what they do to us, much less raised a hand to stop them. From the medical studies that have been conducted, however—both public and private—we know that organochlorines are a universal nightmare, possibly the most destructive thing we humans have yet conceived to do to ourselves, our progeny, and our planet. Already it has been confirmed that chlorinated industrial chemicals cause awful things to happen inside us: cancers throughout the body, birth defects, genetic mutations, suppression of the immune system, impaired reproduction and childhood development, infertility, hormonal changes, and toxicity to the brain, liver, kidney, skin, and other organs.

Thousands of these compounds, all manmade and most of them

highly toxic even in the tiniest of doses, persist in the environment for decades, many for centuries. They readily enter the food chain, working their way quickly from the simplest organism to the most complex, concentrating in the fatty tissues of animals, and becoming more concentrated and potent as one animal eats another, which brings them to us: 177 organochlorines have been identified in the fat, blood, mother's milk, semen, and breath of us humans, and hundreds more are known to be present, but are yet to be specifically identified.

"Shoot," you say, "I've been eating fish from right below a Dow Chemical discharge pipe for twenty years, and I ain't dead yet." But absorbing organochlorines is not like an Agatha Christie character unknowingly taking a teaspoon of barbiturates in warm milk just before bed and then . . . croak. The effects of chlorides in your body may not show up for thirty years. These killers and mutators are never flushed out of our systems, so even infinitesimal amounts—too small themselves to seem harmful—gradually build up to a critical mass that does cause us gruesome harm. Worse, we adults pass them directly to the next generation—they cross through the placenta to embryos, they are sucked up by breast-feeding babies, and they damage our genes, which are then passed along in sperm or ovum to our offspring.

Yet, ludicrously, science, government, industry, and media are paying far more attention to the "crisis" of computer viruses than to the awesomely destructive power of these invasive human poisons. It is a case of willful stupidity, and it proves again that if our national leadership had a brain cell, the poor thing would die of loneliness.

Do you know someone with breast cancer? Who doesn't? It kills 50,000 of our mothers, daughters, sisters, and friends in the United States each year—more than half a million worldwide. The disease is increasing rapidly in practically every industrialized nation and is predicted to top a million deaths a year by 2000, making it the number one killer of women. Little-Reported Fact: People with breast cancer tend to have very high levels of organochlorines in their bodies. Conversely, women exposed to unusually high levels of organochlorines (such as chemical workers, farm workers, and professional golfers) tend to have unusually high rates of breast cancer.

Anyone with an IQ higher than sushi might take these interesting juxtapositions as like, uh, well, you know, like *a clue* that maybe, just maybe, the world production of a couple of billion tons of organochlorines over the past fifty years has something to do with the world's 60 percent increase in the number of breast cancer cases in that same period. After all, a standard rule-of-thumb that doctors use in diagnosing diseases is: "When you hear hoofbeats, think horses."

Women are being trampled by a thundering stampede of organochlorines, but the cancer research establishment is refusing to listen to the hoofbeats. Instead the establishment is going around blaming women themselves for breast cancer—"It's your fatty diets," they scold, "or your genes or your estrogen levels"—while warning women to get mammograms early and often ("Early detection is your best protection," chirps one of their ubiquitous ads).

Nothing wrong with early detection, but how about at least an ounce of prevention, a little focus on the real causes of this epidemic? In lucid moments, even the establishment concedes that no more than 50 percent of breast cancer cases (and cancer groups say no more than a third of the cases) can be attributed to diet, family history, and hormones, so why do they refuse to look at the obvious? Because the cancer establishment is beholden to the chemical establishment, and often they are one and the same. Case in point: National Breast Cancer Awareness Month.

Few do-good events in our country are more favorably received by America's corporate chieftains, media moguls, and politicians than BCAM. When October rolls around, the Breast Cancer Awareness organization rolls out a blockbuster saturation-ad campaign imploring women everywhere to "get a mammogram now." This annual October media blitz has become the "official response" by America's powers-that-be to the spreading scourge: Every governor and mayor proclaims "National Mammography Day" (text of proclamation conveniently distributed by BCAM), every local media outlet produces heart-wrenching feature stories about victims of the cancer, the First Lady dutifully implores citizens to embrace the BCAM message, the postal service distributes millions of special breast cancer awareness stamps, politicians of all stripes pop up at cancer hospitals for photo ops where they promise more public funds to find a "cure," companies like Estée Lauder show their solidarity with the awareness movement by handing out hundreds of thousands of pink ribbons to employees and customers (satin, embroidered, or enameled ribbons are available in bulk by ordering directly from BCAM's handy "product catalogue"), dietitians offer low-fat recipes while admonishing women to eat healthy, and there is generally a month-long, tongue-clucking sermon from those-on-high instructing women to be responsible and, well . . . take care of themselves.

Awareness Month, however, offers not one whisper of doubt that anything or anyone other than women themselves and "fate" are responsible for what is happening to them, that there is something out there actually causing this horrible disease to erupt. Peruse all BCAM's expensive and artfully prepared materials, all its promo kits, pamphlets, press releases, posters, radio spots, print ads, and videos, and you will find no suggestion that there is a manmade chemical connection to what is happening to women.

To understand this studied silence, follow the money.

Breast Cancer Awareness Month is a front that was conceived, funded, and launched in 1985 by a British conglomerate with a name that could come straight out of a *Batman* comic book: Imperial Chemical Industries. But this $14-billion-a-year multinational behemoth is all too real. It is among the world's largest makers of pesticides, plastics, pharmaceuticals, and paper. "Organochlorines R Us" could legitimately be its slogan, though "Pollution R Us" would also fit—one of its Canadian paint subsidiaries, for example, has been held responsible for a third of the toxic chemicals dumped into the St. Lawrence River.

For whatever reason, the conventional media has never questioned where BCAM gets its money, or at least it has never reported that a great big gooey, pulsating conflict of interest is controlling just how little true "awareness" is broadcast during breast cancer's official month. Instead, only those outside the circle of established legitimacy—agitators like Greenpeace and uncompromised investigative reporters like Monte Paulsen of the *Detroit Metro Times*—have had the journalistic gumption to lift the log and look at the uglies squirming under it. It is not a pretty sight, but there it is—as Paulsen wrote in the *Metro Times* back in May 1993: "ICI has been the sole financial sponsor of BCAM since the event's inception. Altogether, the company has spent 'several million dollars' on the project, according to a spokeswoman. *In return, ICI has been allowed to approve—or veto—every poster, pamphlet and advertisement BCAM uses.*"

In 1993 one of ICI's corporate chunks, called Zeneca Group PLC, split off, taking with it the pharmaceutical and ag-chemical divisions of ICI, as well as the Breast Cancer Awareness Month division. Zeneca remains the sole funder of BCAM and retains complete control of its message. Zeneca (which, in a touch of bitter irony for breast cancer sufferers, employs the corporate slogan, "Bringing Ideas to Life"), was one of the gooiest parts of ICI. For example, it was named in a 1990 lawsuit by our federal government for dumping DDT and PCBs into the Los Angeles and Long Beach harbors. It is a multibillion-dollar producer of pesticides, including acetochlor, a cancer-causing, chlorine-based herbicide that by itself brings home $300 million a year to the corporation.

It gets gooier. Zeneca's pharmaceutical arm is also the maker of Nolvadex, the leading drug used in breast cancer treatment. Nolvadex is a highly controversial drug—it does not cure existing breast cancer, but it can help stop it from spreading in some women who are diagnosed early; however, it can also *cause* blood clots, uterine cancer, and liver cancer in those who take it. "Hey, life's a bitch," seems to be Zeneca's attitude, as it pockets almost half-a-billion big ones a year in Nolvadex sales.

What a racket this company has going! It makes billions selling industrial organochlorines linked to breast cancer, it finances its BCAM front to divert public attention from cancer *causes* to cancer *detection,* then it sells Nolvadex to those who are detected. Meanwhile half-a-million more women will die of breast cancer this year, and more will die next year and the next, while BCAM gaily distributes pink ribbons.

Organochlorines are not just about women, though. Check it out, guys: You are not half the man your father or grandfather was. I don't mean in character, courage, or achievement—I mean in pure *macho,* the bottom-line essence of ballsiness: your testosterone and sperm levels. Dozens of scientific studies around the world confirm that sperm counts today average half of what they were only forty years ago, and that the younger you are, the lower your count is likely to be. Moreover, since 1950, there has been a tripling in the number of men who have so few sperm that they are rendered infertile. Worse, what few sperm we have left are not in the shape they once were—there has been a dramatic rise in the number of misshapen and "sluggish" sperm. Here is a humbling fact to ponder: The average man now produces fewer normal sperm than a hamster does. Need I point out that hamsters have much smaller testicles than men?

We might wish that this is simply a temporary problem caused by too-tight Jockey shorts or too much time in a hot Jacuzzi (yes, chemical industry scientists actually have been trotted out to advance such "theories"); alas, however, tests have been done and these "lifestyle" factors have zero impact. Scientists free to look for real causes, on the other hand, keep coming to the same conclusion: Exposure to and ingestion of organochlorines—even in minuscule amounts—messes mightily with our reproductive apparatus. Men with low sperm counts have been found in several studies to have significantly higher concentrations of organochlorines in their semen. Also, men exposed to high levels of organochlorines by their jobs have significantly lower sperm counts.

We are not alone in experiencing the damage. From toads to terns, fleas to whales, there are reproductive deformities in all creatures that find themselves mired in a world poisoned with mirex, dioxin, DDT, and all the other cousins of the synthetic-chlorine family. The chemicals not only cause cancer (by the way guys, testicular cancer has tripled in the past fifty years, too), they also act as "endocrine disrupters." Our endocrine system is the one that regulates the flow of our hormones, which in turn control our whole biological process, including the formation of genitals and the desire and ability to "get it on." By "disrupted," we are talking about alligators with missing or half-sized penises; about eagles that no longer soar sexually, having lost any desire to mate and nest; about male rats that become hermaphrodites; and about a startling increase in the number of boys born with undescended testicles, malformed penises, testosterone imbalance, and infertility. Sheesh—and you thought Lorena Bobbitt was scary.

So, fellows, which is of greater concern to you: the condition of your "apparatus," or the whiteness of toilet paper? There *is* a choice, because effective, economically feasible, clean, safe alternatives exist right now for every major use of organochlorines. For example, paper. To get that totally white, whiter-than-white effect that is promoted by most U.S. producers of magazines, tampons, books, cigarette paper, corporate letterhead, coffee filters, government memos, paper towels, academic reports, and other disposables, International Paper, Champion, and nearly every other U.S. paper maker uses chlorine-based compounds to bleach their wood pulp. Unfortunately the industry's "pure white" results in an impure earth, since the process dumps 700 million pounds a year of a thousand different organochlorines into our waters, making paper mills the number one source of organochlorine water contamination in America—a contamination that *cannot* be filtered out. Paper makers use more chlorine compounds than any other industry except PVC plastic manufacturers, and paper mills use them for *no other reason* than to make paper whiter. I have to repeat that fact because, even as I write it, I cannot absorb the bottom-feeding depths of stupidity it entails: Our society is employing some of the most destructive elements yet invented, chemical elements that literally change our biological essence, for no nobler purpose than whitening paper.

Not only do these carcinogenic, mutagenic discharges pollute our water and everything in it, but the process also permeates our paper products with the poisons, many of which can then deliver a tiny toxic hit directly to us consumers every time we consume them (tampons, diapers, polystyrene coffee cups, toilet paper, white coffee filters, and bleached milk cartons are among those known to leach their toxic compounds). Do not think that buying recycled products gets you off the chlorine hook, either, since nearly all of them started as chlorinated paper and still retain the compounds, plus—in an act of willful dopiness—most of them go through a second bleaching just to freshen up the whiteness of the product.

Nor is the contamination over when you toss chlorine-bleached paper away. The compounds leak from landfills into our groundwater or are spewed into our air by waste incinerators. Indeed, the EPA has found that incineration of the products causes a chemical reaction that actually manufactures more dioxin and other organochlorines. Ashes to ashes, dust to dust, and chlorines to us.

It is folly to keep trying to throw this stuff away, because there is no "away." Far better for the industry simply to do away with chlorine bleaching altogether, since alternatives are readily available. One option is to produce off-white paper, as is being done with more and more print and writing papers, or even to shift to brown paper products, as is happening with coffee filters, napkins, and paper towels. But if it is bright, shiny "*white!*" the producers insist on having, they can have it. Several nontoxic pulp bleachers, including oxygen and hydrogen peroxide (yes, the antiseptic your sweet granny put on your childhood cuts), can produce such high brightness levels in paper that you would not notice any diminishing of the whiteness from paper bleached by killer chlorines. This is not a theory but a growing practice. Some forty pulp mills in Europe and Canada, as well as a couple of mavericks in the United States (see box, next page) are producing such paper, and the world demand for their clean products is zooming.

As an added bonus, it is *cheaper* for the pulp industry to use these bleaching alternatives, so paper products need not cost a farthing more once the conversion is made and is in mass production. In other words, *There is no consumer or economic reason to keep using chlorines to make paper.*

Still, the U.S. industry keeps poisoning us, our progeny, and our globe just to whiten paper. Why? I can only refer you to the conclusion drawn by some of America's top academic analysts of managerial behavior (we are talking Ph.D.'s here, so pay attention): The industry is run by boneheads.

Eliminating chlorine bleaching is naturally and vehemently opposed by ICI/Zeneca, Dow, Olin, Occidental, Bayer, and other major multinational makers of these chlorine chemicals. As we know from long experience with outfits like the Mafia and the Medellin drug cartel, no combine that draws billions from distributing deadly substances will let such silly niceties as morality interfere with their profiteering. So, on the question of doing away with chlorines, go ahead and mark down the position of the chemical producers as "Fuck you."

You can also comprehend the bonehead up-the-butt defiance of most American pulp makers, since there is an initial conversion cost

LOUISIANA-PACIFIC TURNS GREEN

What do a group of hang-loose, Northern California surfers have in common with the Brooks Brothers suits in a Fortune 500 corporation like Louisiana-Pacific Inc.?

The Pacific. A few years ago, the water off the California coast at Samoa gave a whole new meaning to the phrase "surf's up," since the ocean there was damn near boiling with acrid chlorine effluent being piped right out of an L-P pulp mill that regularly made the list of "Ten Worst Polluters in the United States." The scrappy surfers and the $3 billion company soon had something else in common: the courtroom. L-P was sued by a host of mad-as-hell wave jockeys for assorted violations of the Clean Water Act—a suit that led to a consent decree in which L-P agreed to pay nearly $3 million in fines and, astonishingly, to stop using chlorine to bleach pulp at the Samoa mill.

While the rest of the industry has defiantly dragged its filthy feet to prevent any real change in its chlorine habit, suddenly here came L-P out of the blue, saying it would go whole hog, converting this mill to a process known as TCF, or Totally Chlorine-Free.

This was no small commitment, since the cleanup and conversion cost L-P about $120 million, since the whole industry was in a price slump at the time, and since technological pioneers always encounter an abundance of kinks, bugs, and unpleasant surprises down the untrodden path. Among the unpleasantness confronted by this North American pioneer were poor-quality pulp in its first runs, a $30 million loss during 1992 and 1993, some layoffs, and a testy relationship with L-P's pulp peers.

By 1994, though, L-P was getting the hang of it, the whiteness and strength of its product were being hailed by customers worldwide, all 225 workers were back on the job, and the Samoa mill had logged a $10 million profit.

L-P's success puts the lie to the petulant insistence by other industry leaders and lobbyists that TCF is nothing but an attempt by enviro-freaks to destroy Free Enterprise and the American

Way of Life. Chlorine-free technology works economically and it definitely works environmentally. In 1995 California health officials were able to lift a ban on eating fish or shellfish caught in the ocean or the bay around L-P's mill. Also, with the mill's dioxins and other killer chlorine pollutants eliminated, it is "cowabunga" time once again for the surfers of Samoa.

The company now is moving to be the first pulp maker in the world to eliminate completely the discharge of any waste water from the bleaching process by installing a state-of-the-art closed-loop recycling system. Already L-P has cut its discharges of bleaching waters by 80 percent, reduced its energy use, eliminated toxic sludge, and dramatically improved worker-safety conditions.

Whether the industry likes it or not, the newly clean waves now being enjoyed by the surfers along the coastline of L-P's Samoa plant are the wave of the future. Forty pulp plants in Scandinavia, Germany, and other European nations have converted to TCF paper, which already has captured a third of the European paper market. Japan is demanding the same level of paper purity, and the demand for TCF paper in our country will rise rapidly, too, as our public becomes more informed about the deadly impact of chlorine contaminants. Although Washington, D.C., has its head in a chlorine cloud (EPA, for example, still refuses to buy any of L-P's clean paper for the agency's internal use, claiming that such a purchase would "send a political message"), such cities as Chicago, Ann Arbor, and Seattle, and such states as Oregon, Vermont, Maine, and Massachusetts have passed ordinances calling for their governmental agencies to purchase no-chlorine paper products.

Industry lobbyists in Washington grump that L-P is trying to be "greener than thou," but as L-P environmental operations manager Kirk Girard states matter-of-factly: "We are paying attention to our customers. With an eye to the regulatory environment and the market, we took what we thought was a shrewd economic decision."

Louisiana-Pacific's leap into eco-friendly technology is the direct result of the company's commitment to green—the spending kind of green.

that will dent their bottom line in the short run. Even though conversion will pay handsomely for them in the longer run, American CEOs are more nearsighted than Mr. Magoo. Do not count on common sense from them, much less vision.

Far less understandable, though, is the shameful pussyfooting of major paper *users*—like the *Time* magazine empire. Time Inc. absolutely gobbles paper, ingesting nearly 50 million pounds of it weekly and regurgitating it in the refined form of *Time, Sports Illustrated, People, Fortune,* and the whole mess of other slick periodicals under its corporate wing. As one of the biggest buyers of paper in the world, the company does not merely have the market clout to reshape the practices of paper producers, but it has used this clout in the past. In 1990, after postal rates jumped for magazines, Time Inc. demanded paper in lighter weights than then existed. Quicker than a finger snap, the industry delivered. In 1994, when Time Inc. took the lead among major publishers in using some recycled paper in several of its magazines, *Time*'s chief paper purchaser proudly boasted in print: "In the past there simply was not enough recycled paper available to do this. But because we are such a large buyer, we were able to help create a significant supply by asking for it."

This ability to make a difference in the marketplace just by asking for it is what prompted Greenpeace in 1991 to press Time Inc. to lead the way in demanding stock that is free of chlorine bleaching. Imagine the astonished gasps of delight in the entire environmental movement when the company publicly replied that it was doing just that: "We at *Time* take this problem seriously," the editors intoned in a special public response to Greenpeace in the magazine's January 20, 1992, issue, "but we are limited by the availability of paper produced without the use of dangerous pollutants. Most of our paper suppliers are far along with their plans to eliminate chlorine-based pulp. We will use this alternative paper as soon as it is practical to do so."

There it was! *Time* would turn the tide. Two thumbs up, three cheers, four aces, high fives all around!

Ultimately though, all we got was the finger. According to the *Wall Street Journal,* the corporate higher-ups in the Time Warner conglomerate squelched the environmental initiative of its editors. Even though the technology to produce totally chlorine-free white paper has long since been available and is being widely used to produce an abundance of this "alternative paper," and despite *Time's* public promise to switch to the process five years and thousands of tons of unnecessary chlorine pollution ago, the industry leader still refuses to get off these toxics.

Not only did Time Inc.'s executives renege, but they also tried to cover it up with a coat of greenwash, pointing out in a self-congratulatory letter to readers in the April 4, 1994, issue that it had asked its suppliers to begin using a "different" kind of chlorine to bleach the pages of *Time*. The company's move, though, is no more effective than a smoker switching from Marlboro to Marlboro Lights in a self-deceiving attempt to prevent cancer. Called chlorine dioxide, this "different" organochlorine compound is just Chlorine Lite—it does produce fewer dioxins, but it still produces a deadly amount of them, and it still results in hundreds of other carcinogenic and mutagenic substances flowing from pulp plants directly into our water. In fact, chlorine dioxide not only contains dioxins, but its use in papermaking causes a chemical reaction that actually *makes more* of them.

Lest you think that Time Inc. could not afford to switch (a claim that the $4-billion-a-year publisher never has made), nontoxic paper today is competitively priced with the poisonous varieties, and such small American glossies as *Scuba Times* now use the clean paper, even though they have none of the deep-pocket resources of the Time Warner conglomerate. (Go to your newsstand and compare the paper quality of *Scuba Times* with *Time*, which you'll find to be equal, then reflect on the fact that the sticker price is the same for both magazines.) Time Inc. has the wherewithal to stop paper companies from killing us, but to date it lacks the courage of its own editors' convictions, cravenly backing off, apparently unwilling to butt heads with chemical and paper companies that also happen to be advertisers.

It should be noted here, in a parenthetical burst of full consumer disclosure, that this very book is printed on paper, and book publishers have hardly been Earth Firsters when it comes to cleaning up their pulp. The book industry, now dominated by such conglomerate owners as Rupert Murdoch (owner of HarperCollins and now owner of a piece of me as author of this tome), Viacom, Bertelsmann, Hearst, and our old friend Time Warner, goes through more than a million tons of paper every year, and practically zilch of it is the clean stuff.

It occurred to me, then, that if I am going to box the ears of Time Inc. for not doing its part to stop this senseless poisoning, surely in the name of journalistic integrity (not exactly a high hurdle), HarperCollins would want to use chlorine-free paper for my book. I unleashed Betsy Moon, my research "department," to determine if such book paper was available. Her inquiries led her to upstate New York and the Lyons Falls Pulp & Paper Company, which has a solid reputation in the environmental community for advocating the use of TCF paper—*Totally* Chlorine-Free. My Department of Betsy asked an LFP&P official if the company could supply a major publisher's needs if the publisher specified it wanted only TCF paper. "Yes." Is the quality of the TCF paper equal to the chemically dosed variety? "Yes." Is the TCF paper price-competitive? "Yes." Has any major New York trade publisher ever requested TCF? "No."

As luck would have it, though, LFP&P is a major supplier to HarperCollins, so my HC man, senior vice president Adrian Zackheim, made a couple of calls and—bingo!—you have in your hands one of the very first trade books from a New York publishing house printed on clean paper. So even if my ideas don't rub off on you, at least no chlorines will, either.

If only *Time*'s executives had the commitment and common sense of Zackheim. Or, for that matter, if only the chief executive of the U.S. of A. had such a standup attitude. Guess who is the biggest buyer of paper in the world? Uncle Sam. Surely the federal government in all its majesty could use the awesome swath it cuts in the paper market to do what needs to be done.

Enter Our Shining Knight of Papyrus, Bill Clinton. In its infancy, the Clinton Administration innocently attempted to do the right thing by switching the kind of paper public agencies buy. The move came in the fall of '93, when VP Al Gore was causing a media flurry with his "reinventing government" shtick. An executive order was drafted that mostly dealt with buying recycled paper, but down in Section 503 was this small sentence: "Each agency shall establish goals for procurement of totally chlorine-free paper products to be achieved by 1995."

Oh, man, did the pulp hit the fan! The Chlorine Chemical

Council, the American Forest and Paper Association, and a whole hornet's nest of CEOs and lobbyists from every global corporation making or using chlorine bleaches swarmed the White House in a fury, and Clinton quickly went over the hill. (Not everyone knows it, but the President is a long-suffering "thixotrophic." This is a pathetic physical condition commonly seen in such substances as mayonnaise and catsup, which exist as solids until a little heat is applied to them, then they turn to liquid.)

On behalf of our children, Clinton had a historic chance to face down the handful of companies that are saturating our globe with these poisons, simply by being firm about the kind of paper government buys with our tax dollars. Instead, the Leader of the Free World let the companies dictate our government's paper purchases. Then to try saving a little environmental face, Clinton resorted to the last refuge of political scoundrels everywhere: He called for a study. As part of legislation to rewrite the Clean Water Act, the President proposed a federal study of the impact of organochlorines on human health. In the spring of '94, though, under fierce lobbying by the industry and with no serious push from the White House, even Clinton's weasel of a study was quietly snuffed out.

Clinton's EPA weaseled out, too. Since the Reagan years, our anti-pollution agency had been studying the problem of chlorine bleaching and considering regulatory solutions. After hemming and hawing throughout Clinton's first term, and after deliberately postponing a decision until after the 1996 presidential election (no need to arouse the citizenry in an election year), the agency finally announced its new regulations on May 20 of this year. Far from outlawing these killers, EPA chose to side with *Time* magazine and the other chlorine polluters, adopting the weakest standards that were under consideration—standards proposed by the industry itself to switch to the Chlorine Lite technologies. Passing up the opportunity to end this totally unnecessary pollution, our environmental "protector" has done nothing but spray perfume on it, allowing industry to continue creating and releasing chlorinated pollutants that will accumulate in the earth's ecosystem and steadily sicken, mutate, and kill millions of us.

As David Foreman, one of the founders of Earth First! puts it, "The Earth is not dying, it's being killed." It is time to confront the industry-imposed assumption that all these compounds are somehow essential and that the role of government is merely to pick technical nits over whether the allowable residue level for chlorine number 6,374 should be 0.10 parts per million or 0.01 parts per million. No level is safe. The goal of public policy should not be to regulate organochlorines, but to replace them, and to do so with all deliberate speed. After all, the chemical is not the end, but the means; it is not the benefit, but a cost. As Seneca, the great Roman author of tragedies, wrote nearly 2,000 years ago: "Our posterity will wonder about our ignorance of things so plain."

COMING CLEAN

What is the "dry" in dry-cleaning?

You get a whiff of it every time you rip off the thin plastic slipcover from your dry-cleaned blouses, suits, and such. That slightly sour, gassy aroma that wafts from them is perc, or, more precisely, perchloroethylene.

This organochlorine-based solvent definitely will get the spaghetti sauce out of your favorite jacket. But, like the Texas officials who had the brilliant idea of importing Brazilian fire ants in the 1950s to try to get rid of cotton boll weevils, the solution can be worse than the problem. With perc, the spaghetti sauce vanishes, but the solution itself can cause central nervous system damage, reproductive disorders, miscarriages, kidney and brain damage, and increased risk of several kinds of cancer.

A little spritz of perc to clean your pants might not seem like much, but there are 27,000 neighborhood cleaners in America cleaning millions of pairs of pants and pumping out 250 million pounds of perc a year. It has become one of the most prevalent toxic contaminants in the air of America's cities. Especially vulnerable are those who work in these establishments and those who live next door to or above them—in New York City, for example, 60 percent of dry-cleaners are on the ground floor of apartment buildings, and studies have found that these apartment dwellers have an increased risk of cancer, thanks to the steady dose of perc in their air.

Do you bring it home in your cleaning? You betcha, albeit in very small amounts. But these small amounts add up, since once perc's chlorine contaminants enter your body, they are never flushed out, building up steadily in your fatty tissues. A 1995 study by Consumers Union finds that just wearing, say, a newly dry-cleaned jacket and blouse every week can cause you to inhale enough perc over the years to make a measurable difference in your likelihood of coming down with cancer.

So do we have a choice between cancer and going around with spaghetti splotches on our clothes? We do, and it is called "wet-cleaning."

This is another back-to-the-future answer, bringing up-to-date the methods we used before our society rushed pell-mell to embrace the "miracle of chemistry." Essentially, wet-cleaning involves (and I apologize here for having to resort to technical jargon): washing clothes. I do not mean just dumping the whole load in a tub with 20-Mule Team Borax, but *customized* washing and drying.

This requires having skilled employees. Contrary to dry-cleaning, in which most of your soiled garments are simply tossed in the perc machine, wet-cleaners closely inspect each piece for the kind of dirt, sweat, and stains it has. Depending on the degree of soiling and the kind of fabric, the employee then applies the appropriate solution from a lineup of cold water, hot air, biodegradable soaps, hydrogen peroxide, organic shampoos, pressurized steam vacuuming, plain old salt, and other nonhazardous cleansers. Various washing techniques are employed, from spot cleaning to handwashing, and different methods of drying also are used, including drying cabinets and old-fashioned hang-drying.

Wait, you say, my clothes have labels that shriek, "dry-clean only"!—what am I to do? Forget it. These labels are no more to be obeyed than mattress tags that warn, "Do not remove under penalty of law." Practically every fabric can be washed by someone trained to do it right. Indeed, the most common reason we take clothes in for cleaning is that they get a tad rank from our own perspiration, and nothing beats water for removing sweat. Your dry-cleaners don't tell you, but they often use water, too, mainly because perc is not good at cleaning sweat, blood, or urine. Professional wet-cleaners not only can get your clothes cleaner, but their hands-on process also leaves your clothing in better shape. As you probably have experienced, dry cleaning can harm fabrics over time, since it repeatedly saturates them with a chemical solvent that takes the natural oils out of fibers, rendering them brittle and lifeless.

Wait again, you say. Doesn't all of this attention to each garment by skilled workers jack up labor costs and raise my price? Yes and no. Wet-cleaning puts a premium on training employees to be knowledgeable about fabrics and highly proficient at handling them. This definitely means better-paid employees—not a bad thing, I say. But your price should not rise, because the cost of buying, installing, and maintaining wet-cleaning equipment is way less than the capital costs of dry-cleaning, and there is no expensive perc to buy or dispose of. A joint industry-government study found that wet-cleaning facilities require 41 percent less capital, and yield 5 percent higher profits and a 78 percent higher return on investment.

Turning conventional wisdom on its head, such wet-cleaners as Ecomat and Aquaclean in Manhattan, the Green Cleaner in Chicago, Carriage Cleaners in Atlanta, Atomic Cleaners in Beaumont, Cleaner by Nature in Santa Monica, and dozens more across the country are substituting skilled labor for capital, and it is paying off all around: The shops are profitable, the product is top-quality, employees make a living wage, prices are competitive, customer satisfaction is high—and there is *zero* perc in the air.

5

POLITICS

YOU DON'T HAVE TO BE IN *WHO'S WHO* TO KNOW WHAT'S WHAT

- ★ MY DADDY AND OTHER MUTTS
- ★ I HATE LOSING WHEN WE WIN
- ★ POLITICAL MATH
- ★ ADVICE
- ★ CAMPAIGNING
- ★ CHURCH
- ★ THE DAY JIM BOB DISCOVERED DEMOCRACY
- ★ LOOK UP
- ★ DADDY'S PHILOSOPHY

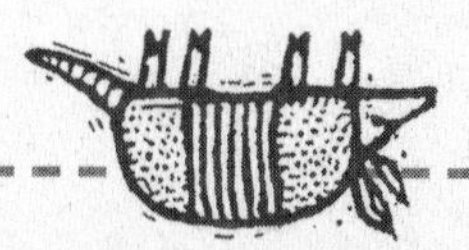

MY DADDY AND OTHER MUTTS

Several years ago I boarded a plane at New York's LaGuardia airport to fly to Dallas. We were taxiing into position for takeoff when the plane abruptly pulled off the runway. The pilot popped on the intercom to tell us—in his best good ol' boy, captain's voice—that everything was going to be okey-dokey: "Folks, ah, our gauge up here shows that our number three engine is out, so, ah, we're just gonna taxi her back to the gate and have 'em replace the gauge. Shouldn't take more than a jiffy, so, ah, sit back and we'll have you on your way real quick-like."

Whoa, did we passengers ever scramble off that plane!

I recall the captain every time I hear some High Priest of the Conventional Wisdom trying to fiddle with our gauges of political reality, hoping to convince folks that everything is running smoothly when we know better. One example of egregious gauge tampering is telling us that the American people are "conservative."

The system—political and economic—counts on your sitting still for this one, meekly accepting that any idea too progressive for, say, Bill Clinton (like a national jobs program, public financing of elections, or single-payer health care for all) is an idea that's simply beyond the fringe, waaaaay too hot for the American people to handle, politically wacky, DOA, a nonstarter, and—as movie mogul Sam Goldwyn once said in dismissing a movie idea—"In two words: Im-possible." The result is a shriveled public debate tolerant only of viewpoints ranging from the conservative corporatism of Clinton's White House to the kooky extremism of Gingrich's Congress.

Never mind that this leaves most of us out, the political cognoscenti are always there to police the boundaries of conventional wisdom. Tune in to any of the political yak shows on TV and inevitably there will be some overpaid and under-IQ-ed peer of the cognoscenti (probably a political pundit or campaign consultant) who'll lean back in full pontifical pose, suck on his teeth for a second, then nail the discussion shut with something like: "You gotta remember, Sam, this is a conservative country." His normally talky colleagues—including the house liberal—will then start to bob their heads up and down in silent tribute to the suffocating power of such gaseous wisdom.

Conservative? Americans? As my ol' daddy would say, that fits about like socks on a rooster.

Take my daddy. For years W. F. "High" Hightower was proprietor and chief BS coordinator of Main Street News, smack-dab in downtown Denison, Texas. He died not long ago, but he was the political experts' quintessential "conservative" demographic: white male, small businessman, solidly middle class, Southern—and while not quite angry, fairly agitated, worried that his America was headed straight to hell in a jet-powered handbasket.

Pollsters could have found "High" most any day with a gaggle of his Main Street cronies telling stories, laughing, and solving the world's problems as they took a break around the Dr. Pepper cold-drink box in his store ("coldest sodas in the Greater Denison Metroplex," he constantly boasted).

"Do you consider yourselves: □Liberal, or □Conservative," the pollsters would have asked these world problem solvers. Sure enough, given the choice, Daddy and each of the others would have responded, "Better mark me down a conservative."

But if the pollsters were to stay awhile, pop a Dr. Pepper, set aside their computer printouts, and just get Daddy and group talking, they'd hear about how out-of-state bank-holding giants are coldly squeezing the life out of little guys like them and little towns like Denison; about how both political parties have become whores to the Wall Street crowd and don't really give a rat's ass about Main Street folks; about how the tax laws are written by and for big corporations and the privileged at the expense of the working man and woman; about how the oil and chemical companies are run by a gang of greedy polluters who'd just as soon piss in your Dr. Pepper as say "hidy" to you; about how . . . well, about how they just don't fit in those tight little squares the system draws up for them, and how they yearn for a political movement, candidate, party, something, anything that steps outside the conventional wisdom and speaks to them, involves them.

No matter what the pollsters, pundits, and politicos say, Americans are no more right-wing than they are left-wing. As the dust bowl troubadour and rabble rouser Woody Guthrie once said, "Left wing, right wing, chicken wing, it's all the same to me," adding that "I've never been a Communist but I have been in the red most of my life." Most of us are mavericks, political mutts—each one of us a heady and sometimes hot mix of liberalisms, conservatisms and (watch out now!) radicalisms.

Yes, radicalisms—at least as radical as the antiestablishmentarianism that is central to the very creation of America; at least as radical, too, as these core political values that most of us hold dear in our hearts: economic fairness, social justice, equal opportunity for all.

You want radical? Here's the radical reality that the powers-that-be don't want you thinking about: America's true political spectrum does not run right-to-left, but *top-to-bottom*. This is the reality the workaday majority of Americans like my daddy actually live. Right-to-left is theory; top-to-bottom is experience. And most people today know they're no longer even in shouting distance of the powers at the top, whether those powers wear the mask of Republicans or Democrats, conservatives or liberals.

One more Daddy story.

Every city and town has its own little power elite—the bankers, real estate barons, old families with parks named after them, that ilk. The "Downtown Crowd," Daddy and his buddies called them. They run things and have an annoying way of setting themselves above and apart from the small merchants, deliverymen, clerks, working stiffs, and the like who came into Main Street News.

At the pinnacle of Denison's power pyramid were the Munsons—old money, big houses, banks, Munson Park right in the center of town, kids sent off to faraway prep schools—"special" people, and they considered themselves as such. As Daddy once said of the family's reigning patriarch at the time, Ben Munson III (or "Ben Three Slashes," as the Main Street News bunch referred to him): "He could strut sitting down."

Well, in 1982, the Munsons tried to strut their stuff in a way that really, really, really PO'd my old man. I was running for statewide office that year and my parents were handling the campaign on the homefront. I was a total political outsider, making a shoestring, uphill race in the Democratic primary against the do-nothing goofball of an incumbent then sitting in the office of Texas agriculture commissioner.

Now, Do-Nothing Goofballism doesn't necessarily work against you in Texas politics—to the contrary, many a Texas governor has risen to power precisely because he fit so snugly in the DNG profile. The big-money interests are particularly partial to public officials who demonstrate an aptitude for creative do-nothingism, and true to form, my opponent had the full embrace of the state's Old Guard, including the Munsons.

They didn't just quietly support him, either. They brought him into my hometown, gave him money, held a Downtown Crowd reception for him, paraded him about like a monkey on a leash, had their pictures made with the grinning fool and—get this—placed that photo right in the middle of the front page of the *Denison Herald*.

Oh the outrage! The insult. The hurt. They might as well have poked Daddy in the eye with a sharp stick.

But this time, there was balm in Gilead. Lo and behold, Election Day rolled around—and I won. Toward midnight, with my statewide victory confirmed, I called home from my Austin headquarters to let my folks know, hoping I wouldn't wake them. Wake them? In the background as Momma answered, I heard the biggest, noisiest party I'd ever known them to have. Not only had I carried Texas, I took 65 percent in Denison.

When Daddy got on the phone, he didn't say, "Way to go, boy," or inquire at all about how I ran statewide. He just kept exulting: "We beat the Munsons, we beat the Munsons, we beat . . . "

That's what people want out of politics—to beat the "Munsons," to have a political movement that gives them a way to go head-on at the powerful, to take America back from the bankers and bosses, the big shots and bastards who are running things and, in the process, running roughshod over them, ruining their sense of what America is supposed to be.

Where is that politics?

I HATE LOSING WHEN WE WIN

In the middle of Florida, just west of Orlando, lies Lake Apopka. Some years ago it suffered a devastating spill of a pesticide containing the dreaded DDT—dreaded because, among other horrors, it messes up our sex hormones something awful. How awful? Ask the alligators in this lake of woes. Male gators there now have testosterone levels so low that many are sterile, and the few newborn males in the lake have penises only a quarter the normal size.

So let's ask the obvious political question: Have national Democratic Party leaders been drinking from Lake Apopka?

How else are we to explain the astonishing wussiness of political leaders so unwilling to embrace their own party's ballsy principles and earthy constituency?

Bill Clinton and his self-proclaimed "New Democrats" are an embarrassment to the party of Jefferson and Jackson, Roosevelt and Truman, JFK and LBJ. Instead of standing proudly on the egalitarian principles of the party and fighting passionately for America's working-class majority, they cling to the moneyed elite and feebly attempt to govern as a bunch of Casper Milquetoasts skittering toward some mythical political "vital center" where they think they'll be safe. On all things they seek to portray themselves as moderates, apparently believing that folks down at the Chat & Chew Cafe will see this as some big political plus: "Hey Earline, I'm all pumped up about the Democrats being so moderate toward the global scumbags and scavengers who're hauling every last one of our good jobs off to the Sultanate of Southeast Shish Kebob, aren't you?"

People want moderates about like a hungry family wants a lecture on diet.

Take Texas, where I ran as a Democrat in seven primaries and general elections during the eighties, winning five of those. I campaigned in every town that had a zip code, and I can tell you there is no place in our state called Moderation. We have New Deal, Progresso, Fair Play, Roosevelt, Gun Barrel City, and Buck Naked . . . but no Moderation. Hell, Texans are a people who have concocted at least two dozen different kinds of hot sauces just to put on breakfast eggs—call it what you will, but this definitely is *not* a moderate culture. We like our food, our fun, our love affairs, and our politicos to be somewhere between plenty-spicy and hotter-than-high-school-love.

My first grab for the gusto of Texas politics came in 1980, when I ran for an oddity of an office called the Texas Railroad Commission. Despite its quaint name, this three-member outfit has zilch to do with railroads and much to do with regulating the powerful oil industry and utilities. I ran as a consumer advocate, making me about as welcome as a tornado in a trailer park.

One day as I was campaigning in Tyler (supposedly a bastion of East Texas business-minded conservatism), I was escorted by a local supporter into the office of a county judge who was one of the area's power brokers. "Be careful here," my escort cautioned. "This ol' boy is terribly conservative, so don't dump your whole load on him."

The judge waved me in, leaned way back in his big swivel chair, propped his feet on the edge of his long desk, clasped his hands over his chest, and almost seemed to drift off as I made my newly modulated pitch on why Texas needed me on the Railroad Commission. When I got to the punch line—usually a fulsome tirade about how the incumbents are nothing but a pack of egg-sucking dogs running side by side with the thieves they're supposed to be watching—I went all mealy-mouthed and mumbled that maybe some of the commissioners were acting kind of in a somewhat biased fashion against us consumers.

You would've thought I'd fired a pistol in the room, that's how fast his feet dropped to the floor, his belly lurched forward over the desk, and his eyes suddenly locked hard on mine. I thought, Oh Jesus, I've hit his trip wire; even mealy-mouth is too strong for this guy. But uh-uh. He squinted his eyes at me and in a gravelly, tight-jawed voice he asked: "Hightower, in your private moments, wouldn't you say they're fucking us?"

Well, yes, Judge, I would, and they are.

This political epiphany in Tyler, Texas, convinced me for good that "Big-D" Democrats make a miserable mistake anytime they back off the "little-d" democratic truth and underestimate the savvy and inherent populism of people. But then I'm not a "New Democrat."

Enter the crowning achievement of the Nouveau Demos: Bill

Clinton, the hail-and-well-met Oxford fellow who bared his inner core to us when he explained he didn't inhale.

Guffaws all around for that. Rush Limbaugh was sputtering and frothing so much I thought for sure he finally was going to implode. But was Bill's claim the obvious fib it appeared to be? I think not. I think this was Clinton in a moment of candor: He really *hadn't* inhaled! Even as a student he wanted to have it both ways (the distinguishing mark of "New Democrats"), seeming to have taken a toke . . . but not really. Indeed, this incident was a harbinger of Clinton's presidency—we would get whiffs of progressiveness, but we would never get high.

He first came into office by talking about "good jobs at good wages," noting correctly that America needs a serious jobs program, a major raise in the minimum wage, an industrial policy, and labor law reform. Instead he gave us the Earned Income Tax Credit to subsidize McDonald's and other low-wage employers, and NAFTA to help Wall Street move more of our good jobs to Mexico. Likewise, his promise of "universal health care for all" became a managed-care plan to fatten up Prudential and the other insurance giants. And now Clinton even justifies sticking a knife in the back of middle-class Medicare families by claiming his stab is not quite as deep as Newt's. Same half-assed performance on the environment, education, tax reform, poverty, and other gut-level issues affecting ordinary folks—compromise, accommodate, capitulate, and back up on the principles and people of his own party.

Reviewing all this, a friend of mine, a Nebraska farmer and "old" Democrat, sighed, shook his head, and said: "I don't mind losing when we lose, but I hate losing when we win."

Clinton and his loyalists scoff at my farmer friend, dismissively declaring that he simply isn't being realistic. It's a "new day," a "global economy," and Democrats "have to march to a new beat." Besides, they privately whine, it's *hard* going against the lobbyists, the Republican reactionaries, and those Democratic Boll Weevils in Congress—if we were to put forward a true Democratic agenda in Washington, why . . . why . . . and their voices trail off in surrender.

Sheesh. If the meek ever inherit the earth, the Clintonites are going to be land barons.

Early in 1998, Clinton momentarily surprised me by declaring in

Rooseveltian terms that he intended to "achieve a landmark for our generation: a social security system that is strong in the twenty-first century." At last, the battle for the middle class would be engaged! A Democratic president was going to throw down the gauntlet to the Republicans, who want to convert social security to a Wall Street lottery based on privatizing and profiteering. What was the President's countervailing proposal? Oh, well, gosh, you see, it's just that, uh . . . He didn't have one. Clinton said that if he (the most powerful officeholder in the world) proposed a specific plan, it would be attacked by the congressional Republicans and others, and he didn't want to get into something like that. So how was he going to achieve this landmark for our generation? "By conducting nonpartisan forums in every region of the country."

Can you imagine Lyndon Johnson or Harry Truman or Franklin Roosevelt brewing such weak coffee? They would pop Newt Gingrich right in the snout, plant the Democratic flag on the highest hill they could find so people could see it, and rally that Nebraska farmer and the rest of us to battle. But the only time Clinton has taken an uncompromising damn-the-torpedoes stand and actually used all the knobs, levers, and gizmos at a President's disposal was to win two items at the top of the *Republican* agenda: NAFTA and GATT. This produced warm applause from the Gucci-and-Pucci crowd, but left the Kmart set real cold.

Sometimes you have to wonder if the chuckleheads running this "new" party could count to twenty with their shoes off. They run a Republican Lite agenda, then wonder why Kmarters aren't voting or are chasing after Ross Perot, Pat Buchanan, or other weirdnesses. They then hire a new set of even-more-Republican-like White House assistants to give them advice on what they should do next, which is always to move closer to the Republicans and to Wall Street.

The Democratic Party needs to move to the right like a tomcat needs to buy a tuxedo. They don't need to move to the "left" either. What they desperately need to do is to move out to the Kmarts. They should take off their natty Brooks Brothers attire, put on those rumpled democratic work clothes they once wore, and begin siding unequivocally with America's working-class families. In other words, *Be Democrats.*

If they would, they would tap into the grassroots power of the workaday majority and not only win, but be able to govern as Democrats. There's an army waiting, but these "New Democrats" won't move it. They remind me of a predicament faced by the last good Republican President America had, Abraham Lincoln.

For a large part of the Civil War, Lincoln could not find a top military commander capable of getting the job done. In April 1862 the President was perplexed to the point of total exasperation by the bewildering reticence of General George McClellan. At the time McClellan's Army of the Potomac had the Confederate forces outnumbered, outgunned, and outmaneuvered in Virginia, yet the general simply would not move his troops against the vulnerable enemy. It was a decisive victory waiting to happen, and Lincoln knew it, but McClellan dawdled and diddled for days. All the while Lincoln was dispatching message after message asking, demanding, ordering, beseeching his general to attack. But nothing. The head of the army would not budge. Finally, after the Confederacy had been allowed to slip away, averting certain defeat, the President sent a final letter dismissing McClellan from his command, saying: "If you don't want to use the army, I should like to borrow it for a while."

If Democrats don't begin using their army—and soon—someone is going to borrow it, and they're not likely to return it.

POLITICAL MATH

"Well, Hightower, if the people have such a hellacious progressive streak running through them, why'd they vote for Newt Gingrich and his gang of right-wing marauders in the 1994 congressional elections and force Clinton to run on a Republican agenda in '96?"

Glad you asked.

They didn't.

Yes, our Loudspeaker of the House was lifted to the dais by voters who gave 53 percent of the congressional seats to Republicans in the '94 elections, putting them in the majority for the first time in forty years. And yes, this electoral turnover led the ever-modest statesman from Marietta, Georgia, to proclaim himself a visionary revolutionist whose ascension to power amounted to nothing less than a Historic Mandate from the People, empowering him to "renew American civilization."

But for all of his huffing and puffing and claiming to be the "Peoples' Choice," Newt is just another politician sipping his own bathwater and declaring it champagne.

The media mostly swallowed it, too: "Public Swings Hard to the Right," blared headlines and newscasts. Even the White House swallowed it, with Clinton himself looking all hangdog after the election and declaring somberly that he had gotten the message from the people and would move to the right, more in step with the country's obvious conservative bent.

Hold it right there! The "people" hadn't voted for Newt, or for anyone else. Amazing Political Statistic Conveniently Ignored by the Media and the Political Establishment: *Six out of ten* Americans eligible to vote in the '94 congressional elections did not do so. Six out of ten. They looked at the choices and said, "Mmmmmm . . . no thanks."

There's the real mandate—a 60 percent majority for "none of the above."

Of the four out of ten people who did vote, two and a fraction marked the Republican box. But half of those told exit pollsters they were not voting *for* the Republicans (much less for Newt's "Contract with America"), they were voting *against* the Democrats. So the

GOP's claim to the mantle of power came down to only one out of ten eligible voters—the same 10 percent minority of true believers, party activists, corporatists, and geeks in golf pants (as Garrison Keillor calls them) that always vote Republican.

The story (and lesson) of the '94 elections is not that the public swung to the right, but that it took a walk.

Which brings us to the last presidential campaign, a total sleepwalk for most voters. By '96 Newt had become so unpopular with the public that his own party put a sock in his mouth and locked him in the basement until the election was over. Newt had leaped to power before people really knew anything about him, and when they finally did get a good look at him in action in '95 . . . well, as the saying goes, the higher the monkey climbs, the more you see of its ugly side.

Meanwhile Clinton spent '95 and '96 bashing welfare mothers, hyping foreign trade deals, demanding more death penalties, siding with industry over the environment, going golfing with corporate lobbyists, and generally positioning himself as a sort of Eisenhower Republican, only not so leftish. He even ensconced Dick Morris, a card-carrying GOP operative (Republican senate leader Trent Lott having been one prominent client of his), as designer and director of his presidential reelection bid, though Morris had to leave at mid-campaign after getting caught sucking a prostitute's toes while she read advance copies of speeches and memos he had written for Clinton.

The actual Republican nominee, "Beltway Bob" Dole, was more boring than oatmeal on a stick. This longtime Washington hack, whose political idol was Richard Nixon, was so inept a presidential candidate that he could not even come up with a campaign theme song. First he tried "Dole Man," done to the tune of the old Sam and Dave classic "Soul Man," but he was threatened with a lawsuit, so he jumped to "Born in the USA," by Bruce Springsteen. The Boss, though, let the media know that he was "not a supporter of the Republican ticket" and that The Bob was using the song without permission. Dole finally settled for "American Boy" by Republican songman Eddie Rabbitt, best known for his hit "I Love a Rainy Night." At least this was better than the music Dole heard as he took the stage at one late campaign appearance: the theme song to "Mission Impossible."

Dole spent most of his hapless campaign complaining publicly that Clinton was stealing the Republican agenda from him. True, but it was not exactly a burning, gut-level issue with potential voters, most of whom were wondering why there wasn't a Democrat in the contest. In a last grasp toward the end of the race, the Republican nominee tried to rouse public anger over corrupt campaign funds flowing into Clinton's White House. Good message, wrong messenger—Bob himself was a notorious old Washington hustler whose various campaigns had collected more special-interest cash and done more favors for contributors in his thirty-five years in Washington than Bill could imagine. There was a surreal moment late in the campaign, like something out of a Fellini film, when a bleary-eyed Dole was pounding the podium in exasperation at a Dallas rally, demanding to know, "Where's the outrage? Where's the outrage? Where's the outrage?" His antimoney pitch might have gotten a better response had his audience not been a crowd of frat boys and sorority girls at the exclusive Southern Methodist University—a group whose well-heeled daddies are among the bigger corrupters of American politics.

During this pathetic excuse of a presidential campaign, I was on the air daily with my call-in radio show, and I found that folks were not merely turned off by candidates so blatantly mired in special-interest money and media posturing, but were fundamentally embarrassed that this is what our politics has become. With no effective opposition, Clinton won, then he claimed the election as a mandate for bipartisan cooperation on a minimalist agenda of Republicanism, and he promptly signaled his willingness to compromise away what few Democratic principles he had not already scuttled to get elected.

As in 1994, however, the only real mandate was for "a pox on both of their houses." In the race for the highest office in our land, when deep economic and social divisions were rending our national unity, the two major party offerings were so obviously inadequate that *more than half* of the electorate—100 million Americans—stayed home rather than sanctioning either of these bamboozlers with their ballots. This was the lowest turnout-percentage of voters since Calvin Coolidge and John Davis excited the people so much back in 1924. Clinton was so weak that even against the comically incompetent Bob Dole he couldn't muster a majority of the minority that did vote. Clinton now avers that he is the choice of the "vital center," a political artifice his spinmeisters crafted to gloss over the ignominy of receiving only 49 percent of the votes of the 49 percent of the people who voted. In other words, not quite 25 percent of the American electorate voted for Clinton, and at least a third of them confided to exit pollsters that they were really voting *against* Newtism.

The beat goes on. Instead of reaching out to that deep, wide, and expanding pool of disaffected voters who either are not voting or are voting against one party or the other (or both), Clinton continues to govern as a cross-dressing Republican and continues to put at the helm of the Democratic Party people who prize big-money contributions over grassroots party building. In the spring of '98, national party chairman Steven Grossman (a multimillionaire Massachusetts businessman and party fund-raiser) trotted out to declare, "We're back. We are suited up, on the field, and ready for the 1998 battle ahead." What did he mean? Had the Democrats recruited a fresh batch of people-minded congressional candidates, built a solid precinct-by-precinct organization, decided to push a pro-worker agenda that would turn voter yawns into cheers? Grossman didn't mention these essential blocks of party building.

He was referring solely to the fact that Clinton, Gore, and the other money grubbers of the party had raised $12 million in the first quarter of the year. In response, the Republican Party, fresh from having killed campaign finance reform in the Congress, calmly announced that it had raised $18 million in the first quarter. The opening whistle for the congressional elections had not yet blown, and already the Democrats were 6 million points behind, because they are on the wrong field playing the wrong game.

Grossman, Clinton, Gore, and the rest would do well to heed a story that South Dakota Senator Tom Daschle tells on himself. The Democratic senator says that during a visit to a nursing home in his state, he walked into the recreation room and found a group of very aged gentlemen whiling away the time watching television. He stepped over to one of them, began shaking his hand, and asked, "Do you know who I am?" The old fellow said, "No, but if you go to the front desk, they'll tell you."

The good news is that some Democrats know who they are and

are trying to reestablish the party's identity as champion of the middle class, not by running slick image ads on television, but simply by going out to the people and visiting with them. For example, throughout 1998, Representative David Bonior, Marcy Kaptur, Bart Stupak, John Lewis, Bill Delahunt, Karen Thurman, George Miller, Bob Filner, and a few other insurgent Democrats still in possession of their democratic ideals have been making a series of two- to three-day bus trips to places like Columbus, Georgia; south central Los Angeles; Gainesville, Florida; suburban Atlanta; and Merced, California. They are meeting not with lobbyists, contributors, or organizational bigwigs, but with folks, hearing the personal stories about how the American Dream itself is being trampled by the Wall Street–Washington stampede to maximize global greed. As mentioned in the introduction to this volume, I was aboard the bus for many of these trips and had the chance to have my own sideline conversations with workers, farmers, and others we encountered along the way. At every stop, someone would finally get around to asking—either in a whispered, quizzical tone or in a bold, demanding manner—the question that bothers a lot of Americans: "Where the hell has the Democratic Party been?"

Sadly, most of the party's leaders remain inside the Beltway, tethered to their corporate funders, lip-synching "free-trade" platitudes with Republican leaders. Representative Cal Dooley, co-chair of the Republicanesque "New Democrat Coalition" in Congress, even expressed dismay that Bonior and his breakaway bunch of *fair*-trade Democrats were going out among the rabble to discuss this explosive issue: "We just spent two days at Wintergreen [a Virginia resort] talking about issues that unify the party," he moaned. "I find this entirely inconsistent with efforts to maintain Democratic Party unity."

Excuse me Cal, I know that I'm not exactly bent over double with intellect, but common sense tells me there is a brighter Democratic future in appealing to the disaffected majority than in constantly trying to out-Republican the Republicans to squeeze a victory out of the voting minority. Some say we need a third party. I say we need a second one.

ADVICE

After I first took office as Texas agriculture commissioner in 1983—having run a headbutting, bloodletting campaign against the chemical companies, the right-wing Farm Bureau, and other forces of ignorance and arrogance—these same forces sent their lobbyists around to see me, offering big toothy smiles, cooing words of congratulations, and proffering checks to help pay off my campaign debt. It's a ritual in Texas politics called "getting well."

The Farm Bureau lobbyist, a particularly smarmy character, took me by the elbow, pulled me closer to him than I ever wanted to be, and with his hot, Binaca-tinged lobbyist's breath whispered confidentially into my ear: "Now that you're in, Hightower, we can start with a clean slate. You move over to the middle of the road and we'll get along juuussst fine."

Later I was laughing about this with a farmer I knew from Deaf Smith County. This corn farmer looked me in the eye and gave me the truest political advice I ever got: "The hell with the middle of the road," he said, "there's nothing in the middle of the road but yellow stripes and dead armadillos."

I wish we had more politicians willing to live by that creed, willing to stand flat-footed for what and who they really believe in. Instead we mostly get tap dancers: "I'm foursquare for the working people, as you know . . . but (tappity-tappity, tap-tap), let's not do anything that might upset the bond market; I'm 100 percent pure-green for the environment . . . but (tappity-tap, tappity tap-tap) we can't do anything that might pinch the profits of the chemical boys, you understand?"

Yeah, we understand you're not going to do diddly for working people and the environment.

I learned early on that if you're going to do anything in office besides hold it, first you have to make choices about who it is you want to help, then (and most importantly) you need to have the *cojones* to take on the powers that are running over those people. Want to help organic farmers? Better strap on your butt-kicking boots and get ready for the onrush of the pesticide companies, bankers, A&M chemical pushers, and the rest of the ag establishment, including their toadying politicians.

Now don't be clucking your tongue and lecturing on how politics is the "art of compromise." Sure it is—compromise is essential to democracy and all that. But one doesn't come out of the chute compromising; ride that bucking, twisting bull for all you are worth first! Reach your compromises honestly, after you've given it your best shot.

Notice that Ronald Reagan, Newt Gingrich, and other far-right protectors of the powerful have had little queasiness about pushing their agenda: "We're for rich people, now get out of our way, dammit." But middle-of-the-road Democrats are always hemming and hawing from the get-go: "Hey, about that universal-health-care-for-all thing, we were just tossing it out for the hell of it, we know you can't sit still for that, but, seriously, tell us, what do you think you could do, maybe a little more money so a few uninsured children could get some coverage, how about that, huh?" In his 1997 State of the Union peroration, Bill Clinton did a little tap dance on this very topic. In one sentence he expressed his heartfelt concern that "40 million Americans still lack health insurance," then just two sentences later he revealed the depths of his political courage by vowing that, by jiminy, his new budget would propose coverage for "up to five million" of them. It turns out he didn't even deliver on this eighth of a loaf. A year later, 41 million Americans lack health insurance. And he wonders why even Democratic historians already have written him off as a "low average" President, right alongside such giants as Gerald Ford and George Bush, and just above the "failed presidencies" of Herbert Hoover and Richard Nixon.

In surveys of what people want in a politician, at the top of the list is the guts to come right out and say what the hell they think. As a man from Iowa said in a recent survey: "I would like to see a candidate that would just simply state, 'I know everybody is not going to agree with me. If you don't agree with me, don't vote for me. But if you vote for me, I'm going to try to do exactly this.'" Seems to me we would have a stronger political system in our country, a much broader and more honest debate about things people care about, and far more participation in the process if more politicians did as this fellow says: Choose a side and come out fighting.

The shame is not in losing, but in clinging to office, afraid to lose. Politics isn't supposed to be a career anyway.

CAMPAIGNING

About the only actual joy I found in political campaigning is the one thing the polling consultants, fund-raising consultants, media consultants, issue consultants, advertising consultants, makeup consultants, GOTV consultants, and all the other purveyors of political "science" are doing their damnedest to eliminate: campaigning. I don't mean the glazed-eyed glad-handing and formal speechifying, but the simple chance to be among the amazingly diverse people of a community or state in a way few others experience—being invited inside their meeting places, to sit around their kitchen tables, to drink in their juke joints, even to join them in their churches, experiencing the laughter, songs, food, accents, and the whole human richness that is the glory of our country.

Consider food. Any politician who complains about the rubber-chicken circuit is a politician spending way too much time at the hotel banquets of organized interests. Ordinary folks don't eat rubber chicken.

My mouth still waters for the duck gumbo I was served when I stopped by unannounced at the IBEW union hall in 1986 in Bridge City, right in the heart of Texas Cajun country, with the duck so fresh there was buckshot in the gumbo. Then there was a South Texas *pachanga* held in my honor about dusk one evening on a vacant lot next to a Texaco station; I thought I had died and gone to heaven as I wolfed down not one, two, or three, but four of whatever that succulent thing was absolutely oozing juices and flavor from inside those plain flour tortillas, learning only later from my grinning hosts that it was *tripas,* or cow guts. Ever had a "Pearl sandwich"? I did on my way to a Czech-American gathering in East Bernard, Texas, not far from Houston. The "sandwich" is one Pearl beer between two others. You want chicken? Try the ambrosial chicken and dumplings with fresh collard greens and sweet potato pie dished out at the little cafe at Pounds Field, the airport outside Tyler, Texas—surely one of the few airport eateries in America with food that people *want* to eat. They come from miles around to get a bite whether they have a plane to meet or not.

Another virtue of grassroots campaigning is that the experience won't let politicians get away with putting on airs and feeling self-important, because, well, as the bumper sticker plainly states: "Shit Happens." Like the time I spoke to the Denison Lions Club. This was a major moment in my life because my Daddy had been a loyal member there since before I was a Cub Scout. He held every office from "tail twister" to president, and in all those years I had never gotten so much as a glimpse inside their meeting room in the funky Hotel Denison.

So here I was, the commissioner of agriculture for the whole state of Texas, forty-something years old, and invited at last into Daddy's Lions' den. I prepared a hell of a speech and was no more than a fourth deep into its wise words and hilarious stories when, out of the corner of my eye, I caught sight of Beuford Thomas, one of Denison's proud police officers, sort of scuttling up to the head table sideways like a sand crab. He began whispering importantly to one of the official Lions at the end of the table. Try as I might to hold the crowd with my spellbinding talk, Beuford's whispering grew louder and more insistent until every Lion's eye was on the end of the table, rather than on center stage, where my self-assurance had begun to wane, despite the resplendent power tie I had selected particularly for this occasion. There was nothing to do but halt and give the floor over to Beuford, who officiously announced that there was a car double-parked out on Chestnut Street, and if said vehicle was licensed to any Lion in attendance, said Lion should deal with it PDQ or suffer the full consequences of the parking ordinances of the municipality of Denison, Texas.

Having spake, he moved on, leaving me in the shambles of my speech. Politics, at its best, is always slightly humiliating for the politicians, which I think is a pretty healthy thing for a democracy. If nothing else, it forces them to get a sense of humor, and Lord knows we're frightfully short of that.

Humor is so rare in politics these days that I've actually been asked by the *New York Times* where I come up with my stuff, like maybe I'd been sipping some well water no one else in politics knew about. No magic to it—in addition to being born full of bull, having a daddy named "High," and being a faithful reader of the comics and sports pages, I'm on the road and on talk radio a lot, constantly rubbing up against the funny bones of hoi polloi. Like the elderly preacher—had to be eighty—who shuffled up to me after a talk I had given, patted my arm, and said in a confiding voice: "Some of these rich folks today seem to think everything belongs to them and they'll even get to take it with them when they die. But you know what, Mr. Hightower? You don't ever see a hearse pulling a U-Haul."

He smiled a big, gentle smile, like he knew he'd just put that line someplace where it wouldn't die. I've lost that preacher's name, but I've sure spread his jewel of a line all across America. Amen, preacher, and thank you.

I especially enjoy the self-deprecating humor that abounds in my state. Just one quick example: I was invited to a town meeting in Panna Maria, a Polish-American settlement southeast of San Antonio. To put an Austin outsider at ease, the chairwoman started the meeting by telling me not to worry about getting everyone's name straight, most of which didn't include a single vowel. She proceeded to tell a story of a local fellow who had gone to San Antonio for an eye exam. The doctor flicked on the light for the eye chart and asked the guy if he could read the bottom line. "Read it?" he exclaimed. "Hell, I know him!"

There's the rub, I think, in today's politics: Politicians don't "know" the people because they simply aren't much among them. Oh, sure, members of Congress will stage their own town meetings every now and then, carefully stacking the crowd with supporters quick to pounce on any hapless dissenter who actually tries to quiz hizzoner; or some computer-chip employees will be blitzed by a visiting congresswoman, grinning, bubbling, and shaking hands for the benefit of the TV cameras trailing behind her, getting it all on tape for a twenty-second bit on the 6 o'clock news. The "people" are little more than fodder for photo ops, focus groups, speechwriters, and campaign ads—not a body for politicians to mess with, and certainly not for them to listen to.

At least the people in these photo ops actually exist. In 1996 the Clinton campaign ratcheted down the bar for cynical campaigning by having the President feign concern for a person who *was not there*. Radio host Harry Shearer tells of whiling away time in Chicago as he awaited Clinton's arrival at the Democratic convention by watching live, unedited satellite feed of the Democrat's train trek from Washington to the Windy City (surely a case of coals to

Newcastle). Of course, the whole whistlestop journey was a staged event, with advance staffers assembling a mass of bodies at each stop and prompting "spontaneous" crowd responses to the Big Man's arrival, speech, and departure.

With every line and move choreographed, Shearer was surprised by an impromptu move Clinton made at the first trainside speech he saw on the satellite monitor. The President suddenly interrupted his remarks to ask solicitously about the condition of someone in the audience who apparently had been overcome with standing so long in the sun. "You need a doctor?" Clinton asked, full of human concern. Then he says urgently, "We need a doctor over here. We need a doctor, and my medical team."

It was a nice, personal touch, and Shearer was impressed. Until he watched Clinton do the same thing at the next stop. And the next. During nearly every trainside talk, all the way to Chicago, the President of the United States pretended that someone in his audience was sick and that he was the Good Samaritan, dispensing aid. The cameras never turned away from Clinton, so viewers never knew that the victim was not there.

Campaign consultants insist that politicians must spend their time with "people who count"—major contributors, power brokers, key media personalities, consultants (of course!), and . . . well, you know, the *players*. Politicians no longer feel compelled to live in our neighborhoods or send their kids to public schools. They lunch at the Casa de la Maison House (as Calvin Trillin calls such places), not at the Chat & Chew, and they party and play with lobbyists and contributors, not with any of us.

This is why members of Congress don't know the price of a loaf of bread, much less know you. It's also why the ineffable, and now infamous, Representative Frederick Heineman of North Carolina (one of Newt's red-hot Republican freshmen elected in 1994) could say with pursed lips and a straight face that his congressional salary of $133,600 a year plus $50,000 a year he gets in other income doesn't make him rich. "That does not make me middle class. That makes me *lower*-middle class. When I see someone who is making from $300,000 to $750,000 a year, that's middle class. When I see anyone above that, that's upper-middle-class."

If ignorance ever goes to $40 a barrel, I want drilling rights on

Fred's head. Eighty percent of us Americans—*eight out of ten of us*—make less than $50,000. Half of us make less than $30,000. This is where the middle class lives. Come on down and visit sometime! Fred's constituents paid him a visit in '96, not just knocking on his door, but knocking him out of office.

Still, Fred's tethers to reality are not that much looser than those of most high public officials, who continue to live somewhere way out in the ozone, largely uninterested in the travails of ordinary voters, much less the thoughts of disaffected nonvoters. Money and media is all that matters, and if you can't deliver either, you rank just above a head of cabbage on their political-importance meter. I saw this cynical reality in action in the '88 presidential run, when Democratic nominee Michael "The Massachusetts Miracle" Dukakis darted in and out of Austin three or four times. His crack team of self-absorbed political operatives had his visits down to stopwatch precision:

4:00 PM	Wheels down, Austin airport.
4:05 PM	MD greets local/state officials on tarmac.
4:08 PM	POLICE ESCORT. MD and entourage to state Capitol; traffic blocked along route, FULL SPEED authorized.
4:13 PM	Arrive at Capitol for photo-op and rally with state officials and party loyalists. Full media coverage.
4:20 PM	MD speaks (15 minutes).
4:35 PM	MD departs Capitol.
4:45 PM	MD arrives at Headliner's Club for private greet-and-go with party contributors ($5,000 minimum soft-money pledge by invitees).
5:30 PM	POLICE ESCORT. MD and entourage to Austin airport; traffic blocked, FULL SPEED authorized.
5:35 PM	Arrive airport and load.
5:45 PM	Wheels up.

Slam, bam, thank you ma'am, the "Duke" does Austin in under two hours.

Witnessing this technically proficient, quickie campaigning on Dukakis's first trip in late summer, and noting national media reports that the guy was not exactly catching fire with minorities, working stiffs, and others whose votes he would have to count on in November, I dared to propose to the East Coast Einstein running his Texas campaign that on his next trip to Austin, Dukakis extend his stay from an hour and forty-five minutes to a full two hours.

It had occurred to me that while there was no possibility of getting our media-and-money-centric presidential nominee actually to campaign among regular people, it just might be possible to get him to wave to some. Call me a radical lunatic, but it seemed worth the effort, what with the presidency and all at stake.

Here was the nub of my thinking: Dukakis's pell-mell rush from the Austin airport to the capitol (and back) takes him right down Martin Luther King Boulevard and right through East Austin neighborhoods of black, working-class families. OK, the nominee is waaaay too busy to stop for an East Austin rally or neighborhood walk-through, but how about this? Instead of his ripping through the neighborhoods, tucked invisibly inside a dark limousine going sixty mph with sirens screaming . . . slooowwww hiiimmmm dooowwwnnnn, maybe to ten mph, and put the Dukester in a convertible so a goodly number of his own constituency might get a glimpse of him caring enough at least to wave to them.

Well, Dukakis's handlers looked at me as though a cockroach had spoken to them. "No time for that stuff, Hightower. We've got a Houston fund-raiser at seven. Gotta go. We're counting on you state officials to get the vote out here in Austin. See you next trip" . . . and "thlllbbbbttttt," they were gone.

Unfortunately the Dukakis campaign was no aberration. Incumbent senators from California, Florida, New York, Texas, or other big states now typically allocate 80 percent of their available politicking hours not for meeting voters or even giving speeches, but for phoning and meeting privately with the lobbyists and other high rollers who bankroll their reelection campaigns. Their average haul is some $10 million per run, nearly half of which they dump into the production and airing of media spots, about 95 percent of which are either howling attack ads against their opponents or pure puff pieces to exaggerate whatever microscopic speck of virtue the beleaguered media consultant can find buried deep inside the SOB's sorry character.

So more and more people, having been stiffed again and again by the high-tech/low-touch politics of both parties, are dropping out. And you know what? The parties don't care! Not even the Democrats!

You expect the GOP to yearn for the good old days when only white male property owners could vote, but Democrats? Nonvoters are overwhelmingly good Democratic prospects—working stiffs. But the once-proud "Party of the People" shows no interest whatsoever in going after the people.

In my final political race, 1990, I teamed with Texas Land Commissioner Garry Mauro to propose a little democratic idea to all the other Democratic statewide candidates, from governor to state judges. The nub of it was that of the $25 million total all of us would spend in the general election, we should allocate about $750,000 to identify, target, and bring to the polls a big block of *guaranteed Democratic* voters who otherwise would not show up.

These are people who consistently go to the polls for the general election in presidential years and consistently vote for the Democrat, but they don't show up for the general election in such nonpresidential years as 1990—the very year that such scintillating and altogether superb candidates as Mauro and me might need their votes.

They live in strongly Democratic counties where the "real election" in off-years like 1990 is the spring Democratic primary, where such locally important and often hotly contested offices as sheriff, county commissioner, and state rep are up for grabs. Whoever gets the Democratic nomination is usually a slam-dunk to win the general election, so large numbers of these voters don't bother to return to the polls in November.

Stupidly, since we statewide Democratic candidates knew we would win these counties in the fall, we had not been targeting these presidential-year Democratic voters, ignoring the fact that we could be getting a lot more votes there if we tried.

Mauro and I calculated that with even a modest effort to find and talk with these folks in selected precincts, simply telling them that their participation in the '90 fall election would matter very much to us statewide Democrats, at least 300,000 of them would bother. This would have produced an extra turnout of better than 5 percent—more than enough to provide the margin of victory in practically every race, just for asking, just for telling people they really mattered.

But no go.

What?! Why would our gubernatorial nominee Ann Richards and all the other statewide Democratic candidates turn down a 5 percent windfall? Because in big states like ours, where it's hard for any of us to "feel" an election with any sense of security, it's easy to fall into the smothering arms of Campaign Consultants.

And their campaign consultants were like Chechen separatists in their histrionic determination to crush this simple flower of an idea. "Spend $750,000 for 300,000 votes?" they hissed into their candidates' ears, "why, that's $2.50 a vote! Think of all the TV ads that would buy in Texarkana." Knock, knock, we countered, we're talking about 300,000 guaranteed votes for *you,* a cushion of 5 percent plus. Anyone home?

No one was, because our idea was untested, and anyway, everyone knows TV ads win elections. Don't they? That's what the consultants tell us.

But something more basic than spending strategy was at the root of the vehemence against actually trying to enlist more voters in this campaign. In an unguarded moment one of the top consultants, a twenty-year veteran in the system, snapped: "We already have enough voters. We don't need any more."

On one level he was saying to nonvoters: You don't matter. But on a more commercially crass level he was saying: Hey, I own the computerized, annotated, and cross-indexed list of current November voters, which I get to resell to you dimwitted candidates every four years for about a quarter-million bucks a pop. I do not, however, have a list of these on-and-off voters, and Mauro and Hightower are proposing not only to create it but, Oh Jesus Mary and Joseph, simply to *give* it to the state party for its use in perpetuity! No way in hell am I going to sit still for this.

So the consultants, much more interested in their fees than in 300,000 additional voters, snuffed our proposal.

I lost my election that year by the narrow margin of only 40,000 votes. Four years later, still with no outreach program to enlist those 300,000 or more presidential-year voters, Ann Richards lost her reelection race for governor.

The consultants, including Mr. We-Don't-Need-Any-More-Voters, are still in business, still collecting their fees and dispensing their excellent expertise to a new set of candidates, while voter turnout continues to dwindle and more and more Republicans replace Democrats in state office.

At the national level consultants are even more powerful, with the media now evaluating candidates' chances not on their ideas or grassroots appeal, but on which team of professionals they assemble and the amount of money raised to pay them. These hired guns are contemptuous of issues and voters alike, seeing both merely as so many inputs to be manipulated through a mechanical politics of fund-raising, polling, focus groups, and advertising—all of which they happen to get paid big bucks to do. Instead of real-life, down-to-earth, hands-on politics, the system is giving us "virtual campaigns," making a mockery of our democracy.

CHURCH

One of the strangest, most kamikaze political strategies yet devised is the ongoing campaign by some liberals to get religion out of politics.

Of course, what has the liberals all worked up is the political skullduggery and general slickum of Reverend Pat Robertson's "Christian Coalition," Reverend Jerry Falwell's "Liberty Alliance," and other outcroppings of the "Christian Right." But be careful where you step on this path, liberal friends, because you might get stuck on your own chewing gum.

Let's start with the fact that the church is on the regular itinerary of any halfway progressive Democrat running for office practically anywhere in America. Not the particular shrines of Robertson and Falwell, but such houses of worship as Sacred Heart Catholic Church on the south side of Waco, Texas.

Sacred Heart baptized me in the practice of pulpit politics one searingly hot Sunday afternoon in August 1981. While Waco is renowned as the Buckle on the Bible Belt and home to Baylor University, it also boasts a burgeoning Mexican-American population, and it was a parish of these good people I was invited to address. It was an outdoor occasion, the annual celebration of Sacred Heart's founding, that gave the priest the excuse he needed to introduce a favored politico to his flock . . . in a *totally* nonpartisan, nonpolitical way, don't you know! I remember it fondly not only because it was my first church gig, but especially because when I arrived on this dry, scorching day and was escorted into the rectory, the first sound that greeted me as I opened the door was the pulsating, high-pitched b-zzzzzzttt b-zzzzzzttt b-zzzzzzttt of a well-used Waring blender, with my priest friend bent over it industriously concocting frozen margaritas and gaily waving me in with the priestly salutation: "Strawberry or watermelon?" Church politicking ain't gonna be half-bad, I decided right then and there.

Some of the richest, most joyous experiences I have had in politics have been in church, including the hour I spent in Reverend Edwin Davis's Galilee Baptist Church in the Acres Home section of Houston. It was October 1986, and I was making a series of scheduled campaign stops at African-American churches. This was always festive and fun campaigning for me. I would be escorted by a respected church member into one Sunday service after another, where I would be welcomed by the minister and allowed to say a few words about the theological imperative of the entire congregation turning out on Election Day to vote for me and the whole Democratic ticket, all according to God's own Cosmic Plan. I would arrive, greet, soak up a bit of the jubilant singing, speak my piece, jump in a car, and dash away to the next church, visiting about four in a single morning.

In these churches, Sunday service runs 11 A.M. to at least 1 P.M., and politicians are supposed to be out of the way by noon, which is when the preacher gets down to the more profound business of saving souls. On this particular Sunday, though, I arrived a bit past 12 at Reverend Davis's church, and he was already wading rhetorically into the cleansing waters of the River Jordan, just beginning to get into the cadence of his sermon. Beulah Sheppard, the church member who escorted me in, gestured for me to sit with her in a pew midway down the aisle and whispered that the reverend would call on me after he was through.

No one told me what a ride I was in for. I thought I had been to church before, but I soon learned that there is church, and then there is *church*.

Pastor Davis is a "singing preacher," literally belting out his homily in song, delivering it *a cappella* in a driving, bluesy, rich, rhythmic baritone that both laughs and cries, is playful one minute and deeply soulful the next, rises to a shout then plunges to a whisper, sweeping you up in its hypnotic musicality, all the while delivering an intoxicating mix of biblical admonitions, everyday anecdotes, twinkling one-liners, personal tragedies, and pointed parables, exploring all nine levels of hell before lifting everyone within sound of his voice onto the true path of redemption and giving us all a teasing glimpse of the glorious light of heaven. He sings extemporaneously for a good forty-five minutes, constantly interacting with the congregation, moving from his pulpit to the pews and back again, never missing a beat. It was the most mesmerizing forty-five minutes I've ever spent anywhere listening to anyone on any subject.

So powerful is this musical messenger, so emotional his delivery, that the experience leaves you all wrung out, and many of the faithful in attendance this Sunday soon were swooning. This was expected, though, and several of the sisters of Galilee Baptist had been positioned to handle the situation. Dressed in white uniforms, they stood watch at the rear of the church, moving swiftly and silently to the sides of those who began to show telltale signs of ecstasy, preventing harm by catching them in mid-swoon, either placing them gently back in their pews or literally carrying them out of the sanctuary into a back room to recuperate.

As the preacher continued, first one, then three, then four, and soon many more of the congregation were hauled out, including several members of the choir. It slowly dawned on me-the-politician that, wait a minute, there are not going to be more than a dozen people left by the time I get up to make my political pitch. Then a truly horrific thought came to me: I'm supposed to follow this preacher! The only white guy in the room is supposed to get up and talk to them about voting for agriculture commissioner after they've just been shown the depths of hell and delivered back to the euphoria of salvation?

But there I was. The time came and the reverend did call me forward. I faced the remaining churchgoers, who looked up at me through blinking, emotionally exhausted eyes, as though I had just shaken them awake from some dream. I stood as close to Reverend Davis as possible and said, in essence, "I'm with him." Shortest speech I ever gave.

I know, I know—for every preacher, priest, or rabbi letting me in the door, there's a Pat Robertson praising God for Jesse Helms, embracing Newt Gingrich's "Contract with America," and waging a terrible Holy War against liberals, environmentalists, women, unions, *Sesame Street,* the PTA, books, welfare recipients, *Time* magazine, UNICEF, Ted Kennedy, Episcopalians, government, Hillary Clinton, Democrats, democrats, and even *The Lion King* (they say that if you look very, very closely, there is a moment in Disney's animated flick where a cloud of dust rises up and seems to spell out S-E-X).

But the presence of a few seriously twisted Beelzebubs in the pulpits of the "Christian Right" doesn't mean everyone in the far-flung flock of conservative Christianity is afflicted with a terminal case of political right-wingitis. After all, we're talking about some 51 million people. Here's my contrarian's notion: Instead of progressives expending energy trying to exorcise the Pat Robertsons from politics, better we should put our energies into winning the political souls of those church members the Robertsons claim to speak for.

Whoa, you say, we might as well try to herd cats as try winning over those people!

But who are "those" people?

In large part they are working class and poor, often they are black and Hispanic. It is not politics that draws them to the churches of the religious right, but the church itself—both the welcoming sociability of the church family and the reassuring religious doctrine that God takes an active role in their everyday lives.

Politically they are hybrids. On such family and cultural issues as abortion, prayer in schools, birth control, homosexuality, and pornography—count them as staunchly conservative. But on economic issues—mark them down as William Jennings Bryan.

In an important, insightful, in-depth, and therefore widely ignored analysis, two Indiana sociology professors reported that Robertson and Falwell try to throw a wide loop with a mighty short rope when they claim that conservative Christians bless the Republican right wing's congressional assault on poor folks, working families, and that old devil government. Yes, Falwell, Robertson, and other political televangelists shout, "Hallelujah" at every whacking of Medicaid and cry, "Thank you Jesus!" when the privileged are awarded a new tax loophole—but if you listen, you won't hear any big chorus of "Amen" being shouted back by the congregation.

Remember, the congregations are working-class families, middle income and less. While they might not love government, they know they don't want to be left to the tender mercies of unbridled big business, which is the secular salvation offered by the Right Reverend Gingrich. Nancy Davis and Robert Robinson, authors of the groundbreaking report, found that rank-and-file religious conservatives "are actually more egalitarian and more favorable toward government intervention to assist the poor and reduce class inequality than are moral progressives." Specifically they found that these supposedly "conservative" churchgoers strongly favor government efforts to reduce the income gap between rich and poor, government providing a job for everyone who wants one, government-paid health care for all, ample welfare spending, aggressive labor unions, making corporations put more of their profits into better wages and benefits for workers, and . . . well, an agenda that is pretty much the opposite of what Republican politicians are pushing and Democratic politicians are meekly accepting these days.

In conclusion, wrote the analysts: "The term 'Religious Right' implies a broad-based conservatism that does not exist."

It does not exist in the Bible, either. At Waples Methodist Church, where I grew up steeped in the blood of the lamb, I recall many a sermon that began with Jesus's words in Matthew 19:24: "It is easier for a camel to go through the eye of a needle than for a rich man to enter the kingdom of God." I recall that the greatest sin of them all, "the root of all evil" according to Jesus, is "the love of money." I recall the vivid and passionate retelling, by Sunday School teachers as well as preachers, of the time Jesus threw the money-changers out of the temple—threw them out for charging interest rates far less than the usurious rates we are now charged on our Visas and MasterCards, by the way. I recall Mary's song in Luke 1:52–53: "He hath put down the mighty from their seats and exalted them of low degree; he hath filled the hungry with good things, and the rich he hath sent empty away." I recall that Jesus fed all the masses that day on the mountainside at the Sea of Galilee, not just those few who could afford the price of fishes and loaves. I recall the hellacious fate of Dives, the rich man who would not give even the crumbs from his table to the beggar Lazarus. And I recall that Jesus flatly said: "You cannot serve both God and Mammon."

Such graphic condemnations of greed, such potent parables conveying what God thinks of the powerful and the privileged, such godly alarm at the disparities between the rich and the rest of us are central to the teachings of the Bible. In the good book, Jesus talks about economic justice more than almost any other subject, and the apostles make clear that the most important ethical/religious test in the Judeo-Christian experience is how we treat the least wealthy, the least important among us.

Yet instead of gleefully grabbing this treasure trove of moral progressivism and reaching out to the millions of our fellow citizens who cradle these biblical lessons inside their hearts, there is an unfortunate tendency among most liberal leaders to view the Bible and its adherents cynically, and they rather prissily reject the emotional power of religion to move people. Instead of appealing to the religious, liberals today are pursuing a self-defeating strategy of trying to ignore or ostracize them.

OK, admittedly there is a chasm wider than Rush Limbaugh's mouth separating liberals from these people on social issues. I've had it explained to me thusly: "Hey, Hightower, the Pope has his mind riveted shut on the choice issue, he's a complete jerk-off on women in the priesthood, he's from Never-Never Land on birth control, and he wears a silly hat . . . so screw him."

But isn't this the same guy who speaks out against income disparities in our country and around the world, who lambastes global corporations for pillaging poor workers and ravaging the environment, and who has even traveled to America to say to our faces that Gingrich's "Contract" is a moral stinker? Isn't there something here—something big—to build on politically?

And even though Pat Robertson is given to such bull-goose crazy crap as referring to non-Christians as termites, declaring that "the time has come for a godly fumigation," ordinary, get-up-and-go-to-church conservative Christians don't agree with that any more than they agree with those letters Robertson sent to members of Congress in 1995, claiming that giving new tax loopholes to wealthy speculators was the Christian thing to do.

When these religious folks sit down at their kitchen tables in the morning for a quick bowl of cornflakes and a cup of coffee before dashing off to their jobs, they're not talking about abortion and prayer in the schools. They're wondering how they can keep making ends meet on the miserable paychecks they're getting, they're worried about paying for Grandma's nursing home bills, they're concerned about whether their daughter will get to go to college, they're discussing the rumors that Dad's company is going to shut down and "outsource" his work to nickel-an-hour villagers in the Indian Punjab, they're realizing they'll never get to retire, they're thinking . . . progressive thoughts, the kind of thoughts Jesus would bless and nurture if he were sitting with them. The fact that these families believe that Jesus *is* sitting with them ought not deter progressive political organizers from asking to pull up a chair, too.

It is only recently that progressives abandoned these kitchens, and we are all the poorer for it. From William Jennings Bryan's "Cross of Gold" speech to Martin Luther King's "I Have a Dream" speech, from Dorothy Day's Catholic Worker's crusade in the cities to Cesar Chavez's United Farm Workers' fight in the fields, great progressive movements in our nation have been unabashedly evangelical, often connected directly to the churches and always rooted deeply in the radical message and democratic values of Jesus. Why give up such moral power, especially at a time when economic fairness and social justice are at the core of so many people's daily struggles?

There is a Christian Progressivism awaiting, not just among a religious intelligentsia or a clique of card-carrying liberal Christians, but among the millions of Pentecostals, Catholics, holiness churches, and other evangelicals wrongly presumed to be the property of the "Christian Right." Despite the self-serving claims of Robertson and other para-political, false-profits of the right wing, most of their own flock cannot find a home in the racist, antiworker, laissez-faire, caveat emptor, never-give-a-sucker-an-even-break, big-business, Republican Party of Newt Gingrich, Phil Gramm, and those other hounds from hell. (While Republicans keep insisting that God is on their side, I cite biblical evidence that Jesus actually is a Democrat—check Matthew 21:1 and you'll find that he didn't ride triumphantly into Jerusalem that first Palm Sunday on any elephant, but on the sturdy back of a donkey. I rest my case.)

Sadly, though, the only Democratic alternative Christian believers see today is the strategically challenged leadership of Bill Clinton and his clueless band of "New Democrats," who are marching off into the distance under the twin banners of "social liberalism" and "economic conservatism"—the exact opposite of what makes this massive constituency tick. These New Democrats are like the guy in a joke I was told recently by my dentist. Seems a guy went to his doctor, complaining of sexual inadequacy. "Nothing wrong with you that a little light exercise won't cure," the doctor assured him. "Tell you what, I want you to begin taking walks. Just walk about five miles a day for the next ten days or so, then call me back and let's see how you're doing."

Ten days later the fellow calls in and says, "Boy, Doc, you were right about this exercise business, I feel great, strong as a bull!"

"So," the doctor asks, "how's your sex life?"

"Sex life?" the guy asks incredulously, "Doc, I'm fifty miles from home."

THE DAY JIM BOB DISCOVERED DEMOCRACY

Senator Mitch McConnell is fast becoming a favorite of mine. At least as smart as Dan Quayle, this Kentucky Republican can always be counted on for a boffo performance whenever the Senate considers any sort of political reform. He's against it, you see, whatever it is.

Like some amiable twit from one of P. G. Wodehouse's send-ups of the British upper crust, my man McConnell simply cannot fathom why there is a need to change anything. For him the system is doing smashingly well, thank you, just dandy. Stop lobbyists from giving all-expenses-paid junkets to senators? Outlaw the "soft-money" loophole that lets a corporation grease the wheels of government by injecting half-a-million dollars (or howsoevermuch it takes) into a political campaign? Mitch doesn't see the point.

Especially mirthful was McConnell's contribution to a heady debate the Senate fell into one day concerning you voters, or more specifically, the majority of you who are choosing not to vote. "Why?" was the question perplexing our learned Solons. Rising to the occasion, Mitch opined that he saw nothing at all to worry about: "Low voter turnout shows that people are happy with the job we're doing in Washington."

Happy? Mitch, if you get one digit dumber, we're going to have to start watering you twice a week.

People are not "happy," nor are they "lazy," "stupid," or "apathetic," three pejoratives often hurled at the nonvoting public. What they are is mightily PO'd, as can be confirmed by anyone who has put his or her finger on the throbbing pulse of the body politic. The Harwood Group has done exactly this regularly and extremely well during the past few years. They're a nonpartisan, independent outfit that has been conducting a series of in-depth and very telling conversations with plain citizens about their attitudes toward our national political system.

Number one finding is that it is no longer "our" system. You don't have to be in *Who's Who* to know what's what, and people do know that there has been a "hostile takeover" of America's electoral and governmental processes by a political class of professional politicians, big-money contributors, lobbyists, and the media. In his latest round of discussions in various towns and cities, Harwood finds that people are not just angry, but "exasperated" by the politicians who, as a Jacksonville man put it, "don't have a clue as to what we're up against as working people."

People feel locked out by a system in which politicians and officials listen to moneyed interests and not the common interest—the same moneyed interests running roughshod over them in the workplace. "The individual is no longer represented," said a man from Laconia, "it's whoever has got the most money." "In order to run for office, you have to be in a certain tax bracket," a Tampa woman complained, adding, "you've got to be dealing with corporate America." "Screw those $500-a-plate dinners," exploded a man from Orlando. "Every one of those candidates should be out at Fleet People's Park or Lake Keola so we can go down there and tell them what we think for free."

At the root of the public's negative feeling is something entirely positive: Americans *want* to be involved in the politics and governing of their country. Contrary to the conventional wisdom, which says the majority is too self-absorbed with their own lives to care about public life, they have a deep longing to participate. "It's not that people no longer have a sense of civic duty," said a Seattle man who was part of Harwood's conversations, "but it's that they don't have a sense of power."

But give us even a tiny opening, a sense that our voices will be heard, a sense that change is possible if we take some action, and . . . well, stand back. Although you do not see it covered on the network news, and while doofuses like Mitch McConnell know less about it than a hog knows about Christmas, America's grassroots are alive with progressive agitation and political participation. Many of the same citizens who have written off national and congressional politics as a "public relations charade" are devoting time and energy to causes and campaigns at the local level where, as one devotee puts it, "politics still has hair on it."

Anyone who thinks people just don't give a damn and won't get involved ought to ask Jim Bob Moffett about that. On the long night of June 7, 1990, democracy rose up in all its chaotic glory inside the city council chambers of Austin, Texas . . . and bit Jim Bob square on the butt.

He's still chapped about it.

Jim Bob Moffett is an actual person. You could look him up in *Business Week* magazine, which listed him in 1997 as America's tenth highest-paid corporate heavy, hauling home $33,732,000 (or $16,800 an hour) for sitting in the big chair of the global conglomerate Freeport McMoRan (Corporate Motto: "If brute force isn't working, you're probably not using enough of it"). A worldwide mining-petrochemical-development octopus that grosses in excess of a billion-and-a-half bucks a year, the conglomerate is involved in digging, spewing, bulldozing, and other activities hither, thither, and yon that have made more than money—they've made some serious environmental messes, ranking Freeport McMoRan more than once as America's number one water polluter. Aw hell, is Jim Bob's attitude, to make lard you've gotta boil a hog—it's a messy process.

Moffett, who is the "Mo" in McMoRan, is a pompadoured rooster of a guy, pretty well satisfied that it's his cock-a-doodle-doo that brings the sun up every morning. So when this native Texan and former University of Texas football letterman strode to the podium in Austin's city council that Thursday in June, every bit as big as his six-feet-four-inches, he fully expected to get what he wanted. After all, did he not enter the chamber with a covey of lobbyists who had been purring sweet nothings into the ears of the council members for weeks? He did. Was it not true that he had put the president of the University of Texas on Freeport's well-paid board of directors? It was. Had he not already donated $2 million to UT's College of Natural Sciences, another two-and-a-half-million to boost the salaries of UT's athletic coaches, and an extra $200,000 to build the Mr. and Mrs. James Robert Moffett Room of the ex-student association's building? He had. Was this deal not wired? It was.

Except for one loose wire he had never contemplated: the people of Austin.

What Jim Bob wanted from the local government was just a few "little ol' technical approvals" to let him build some homes and stuff on the hills that frame Austin's west side. "Some homes"? More than 2,500 of those high-end designer houses and 1,900 upscale apartments. "And stuff"? More than 3 million square feet of office and shopping space, and not one, two, or three, but four eighteen-hole golf courses. Plus thoroughfares, roads, and streets. All on 4,000 acres.

But it wasn't the scale of the development that caught Austin's attention, it was—as real estate agents say—"location." Jim Bob wanted to build his playpen right up against Barton Creek, one of the five creeks that merge just past his proposed Jimbobville and then disappear underground through limestone fissures and openings in the creek beds. Seven miles further down, inside the city of Austin, these cold, crystalline waters reemerge as Barton Springs, burbling out into as sweet a swimming hole as the gods could imagine. The springs are not merely in the center of Austin, they literally are the spiritual heart of Austin.

Here, then, is where J. B. Moffett proposed to take a corporate crap. All the contaminants from his garish development—the spilled gasoline and motor oils, the fertilizers and pesticides running off the lawns and golf courses, the construction solvents, the household and commercial bug sprays, the sewer leaks, the inevitable detritus of such a human and commercial mass—were headed for the creek, the springs, and the heart of Austin.

Until the "Barton Creek Rebellion."

Two weeks before the June 7 vote, word began spreading rapidly through town about this Moffett gentleman and his designs on the springs. His lobbyists reportedly had lined up four sure votes on the seven-member council, including that of the mayor. It did not look good, but a couple of talk-radio hosts and the Austin public radio station began to pound their electronic pulpits and fulminate against the cabal between Jim Bob and the council. The *Austin Chronicle*, a weekly alternative paper, jumped in with journalism at its gritty best, grinding out column inch after column inch loaded with details about this sleazy deal—details that the local establishment paper (whose publisher actually doubled as the president of the chamber of commerce) either didn't bother to dig out, ignored, or intentionally buried.

Then the people took over. Neighborhood and environmental groups began to walk and talk, but the most encouraging sign was that nonactivists were suddenly, spontaneously on the phones, absolutely aghast that the nation's number one water polluter was going to be allowed anywhere near hallowed water. One of them, third-generation Austinite Cathy Lee, told a *Texas Monthly* writer: "I'm not in those environmental organizations, and I didn't want to be aligned with them. We just spread the news by word of mouth. People asked, 'Who are you with?' and I said, 'Myself and Barton Springs.'"

Comes 4 P.M., June 7. The mayor nervously gavels the hearing to order—nervously because every seat in the chamber has been filled for hours, and the suits are badly outnumbered by the T-shirts. The rowdy crowd flows out into the hall, into the street, and on down the block to Liberty Lunch, a classic Austin music venue and bar that has thrown open its doors as a makeshift headquarters for those who have come to fight Jimbobocracy.

Still, Moffett feels good. His team of lawyers, lobbyists, and engineers is headquartered in a Bravo RV parked in the lot right behind the council chamber, a lot usually reserved for council members only. He has hired eight "sitters" to take front-row seats early in the day and hold them till he is ready to make his entrance, which he does just as the mayor's gavel comes down. Then Jim Bob steps up to bat.

His role is to establish his legitimacy in the community and to assure everyone of the environmental sensitivity he and his company will bring to the task at hand. With his first swing, he goes for the fences, trying to wrap himself in the orange and white colors of UT. With heartfelt pride and a booming voice he declares: "I graduated with the highest grades of any football player at the University of Texas." Now this assertion might have won him warm applause at a Rotary Club luncheon, but it was met with hoots and waves of derisive laughter in this chamber. Strike one.

His next swing is a professional stroke, telling the locals that he and his engineers are the finest in the land and know precisely what they are doing: "As a geologist, I assure you I know more about Barton Creek than anybody in this room." Foul ball; more hoots and louder derisive laughter. Strike two.

His final cut is to attest to the integrity of his company and the character of those who run it, adding that none of his esteemed board members, "including Henry Kissinger," would have any part of harming anyone's environmental quality. Great gales of howling derisive laughter for this whiff, especially from those who recall that Dr. Kissinger secretly bombed Cambodia's environment clear to hell. Strike three, and the Mighty Jim Bob has struck out.

Next up were the people. Nine hundred signed in to speak their three-minute pieces. They spoke all night—some heard the hearing being broadcast on radio after work and decided they should speak out, too, so they drove back downtown to the meeting; after the live music bars closed about 2 A.M. the musicians showed up to speak for the springs; others arrived as late as 5 A.M., saying they were on their way to work, but wanted to stop first and support the cause. As the *Chronicle* reported: "One by one, using logic, research, emotion, humor and ridicule, the people of Austin tore the Barton Creek development apart. Poems were composed for the occasion. Videos were shown. Songs were written and sung. People spoke of childhood memories, sacred places, departed loved ones. Eloquence was on parade."

Democracy, too. The last speaker finished around 5:30 A.M., thirteen hours after the hearing began. Still the people stayed, eyes riveted on the council members who were whispering and gesticulating, frantically trying to devise some last-minute compromise language that would give Jim Bob what he wanted but not cause a public hanging of council members right on the spot. A weasel motion was offered, but a council opponent of the deal shredded it for the sellout it was. Then George Humphrey, a council member who had been counted as one of Jim Bob's four votes, spoke: "We've all got to look into our hearts right now. I baptized my baby in Barton Springs about ten months ago, and I want to make sure he can baptize his baby in Barton Springs. I'm asking you to rethink what you're doing because it's wrong, wrong, wrong."

There was a bit more wrangling on the dais but finally, about 6 A.M., the vote was called. Seven to zero for Barton Springs.

This victory is still a long way from assured. His ego stung, Moffett continues wailing and flailing to this day. He threatened the council and the people of Austin, filed lawsuits right and left, forced a second vote at the city council (he lost) and a special election (he lost again), then ran to the state legislature with even more lobbyists in tow, stuffing money into legislators' campaigns and produced a "Bail Out Jim Bob" bill, which Governor Ann Richards vetoed. Eight years later his assault on Austin continues full-tilt, with him trying to defeat all pro-environment council members, hiring still more lobbyists and lawyers, sending threatening letters to journalists and even some professors who oppose his corporate buccaneering—and I have no doubt he is bargaining with the devil himself, trying to undo what democracy wrought back in June 1990.

But the people felt their own power that night, and they have stayed vigilant since. The Jim Bobs will no longer just roll over them.

LOOK UP

I confess to envying the powerful in our society. Not for their exalted status, but for the relative ease they have in making whatever political play strikes their fancy. It is an ease stemming naturally from the twin facts that (1) they are powerful, and (2) they are few: Make a dozen calls to the boys (and they are nearly all boys), touch a few bases, maybe take a meeting or two at the club, and suddenly it's snap, crackle, and pop—money begins to move, PR firms jerk into gear, bills are introduced in city council or Congress, editorials are written, ducks get in a row, skids are greased, and the next thing you know there are odd breakfast conversations all over America:

"Hon, what's GATT?"

"What's a gnat?"

"No, GATT, G-A-T-T."

"I don't know, what is it?"

"Well, I don't know, either, but it says in this little item that Congress just passed it."

On the other hand, on the progressive side of politics, where our forces are short of button-pushing power and long on numbers and diversity, getting everybody organized is about like trying to load frogs in a wheelbarrow.

This democratic handicap is cheerfully exploited by the powers-that-be, which are always Johnny-on-the-spot to spread lies, money, gossip, and whatever it takes to divide us. They say to farmers, "Your problem is with those labor unions"; they say to unions, "The environmentalists are out to get you"; they say to environmentalists, "The minorities are against your plans"; they say to minorities, "It's those gays you need to look out for."

But hold it. It's not gays who are knocking down wages, looting pensions, canceling health plans, and stealing our middle-class aspirations from us. It's not minorities who are using GATT and NAFTA to haul our good jobs out of the country and undermine our environmental and worker-protection standards. It's not environmentalists who are jacking up our interest rates, cutting back on banking services, and taking away all local control over our own money. It's not union members who are depressing agricultural prices, bankrupting good farm families, and closing off the opportunity to farm.

None of us are enemies. We are natural allies—as Jesse Jackson puts it, "We might not have come over in the same boat, but we're in the same boat now." Whenever people realize this they are empowered, for it gives them the political clarity needed to rise above the distrusts and divisions used constantly to separate one group from another, allowing people instead to focus jointly on real enemies. In Oklahoma City in November 1997, I saw this "same boat now" phenomenon at work during an all-too-rare meeting between two groups often pitted against each other: environmentalists and farmers. What brought them together was—excuse my French—hogshit. As discussed in an earlier chapter, a tidal wave of hog excrement is being loosed on rural America as corporations like Seaboard Inc., Land O' Lakes, Nippon, and (I love this name) Pig Improvement Company roam the countryside plopping down dozens of massive and massively polluting hog factories. The weekend-long Oklahoma City meeting brought some 200 citizen activists together from several states to strategize and organize against these swine combines, with roughly half of the attendees in boots, the other half in Birkenstocks. At the Saturday luncheon, where I was to speak, a Western Oklahoma farm activist, Suzette Hatfield, received an award from the Sierra Club for her grassroots leadership on this issue. In her short thank-you to the crowd, she noted that just a year earlier she had been shopping with her daughter in a mall. The girl had come up to her clutching a picture calendar, asking her mother to buy it for her. No way, Suzette told her daughter then—"Not a dime of my money is going to that bunch of environmental radicals." It was a Sierra Club calendar she was saying no to, yet here she was less than a year later accepting an award from those very radicals, saying how proud she was to be teamed up with them in the Oklahoma hog wars.

Pogo, Walt Kelly's cartoon character, is famous for saying "We have met the enemy and he is us." But with due respect to my favorite possum, the enemy is not us people generally, but specifically those in the top suites who lord it over us, our environment, and common sense. That is why we need more of the radicalism of ordinary citizens like Suzette Hatfield who refuse to be lorded over. The central issue that unites us today is this: *Too few people control too much of the money and power in our society,* and they are using that control to take more for themselves at our expense.

It is no longer enough for us to be passively *pro*gressive—we must become *ag*gressive again, because the powers-that-be have become rabidly *re*gressive, reaching for it all, no matter what it costs the majority or what it does to our country.

We should not be glancing side to side at one another, squabbling over this one being gay or that one being in a union, and we damn sure should not be looking down at those below us on the economic ladder, fussing over which one of us got on the ladder first. Rather, we should all look up, keeping our eyes focused on those at the very top who are trying to cut the ladder off for all of us.

DADDY'S PHILOSOPHY

My daddy is buried in the rolling Texas prairie at Denison, with a modest marble headstone to mark his grave. But there is another monument in town with his name on it, too, a couple of miles from the cemetery, where West Woodard Street dead-ends at Maurice. Right there is a Little League baseball field, the first that Denison had, and at the edge of the field is a small sandstone monument with a brass plaque acknowledging the half-dozen men who started Little League ball in our town, including W. F. "High" Hightower.

The marble stone marks his grave, but this stone marks the way he lived and what he valued. Daddy and his Lions Club buddies got the city to donate the vacant lots that now form the ballfield; they organized the teams and came up with a few dollars to buy equipment, uniforms, and such; then they "volunteered" us boys to clear the lots of rocks (approximately 117 billion of them, as I recall) and to do the shoveling, hauling, leveling, planting, and other chores necessary to turn this raw space into a ballpark. Once we made a place to play, Daddy and other folks in town coached the teams, did the official scoring, ran the snowcone stand, and announced the games on the squawking P.A. system.

I played five summers on that field—once batting .517 (a far better year than I have averaged since; little did I know I had peaked at age twelve). Even after I moved on, though, Daddy stayed involved in the Little League program, because it was what the community needed. He was not doing it only for his boys, but for the community.

He believed in that concept. He believed that everyone should pool efforts when the common good calls for it, whether pitching in locally to start a Little League or being taxed nationally to assure everyone's social security. "You're no better than anyone else," he would tell me as a maturing boy, "and they're no better than you. We're all just people and we need each other." Not that he was any kind of philosopher, for God's sake, or any do-gooder, either—just an ordinary American of common sense and goodwill who grew up on a hardscrabble, Depression-era tenant farm and had learned from life that *everyone does better when everyone does better*. That is what passes for political philosophy in Denison, and it remains the best I have found in all my years wandering and studying.

Also I found that his outlook is shared by a majority of folks everywhere, today just as surely as it was in Daddy's time. Yet it is this very concept of the common good that the powerful in our society are rending as fast as their fat little hands can rip.

In the class war of the past couple decades, the corporate and investor elites have made out like bandits by shortchanging the middle class on economic gains that the whole society has produced. They have forcibly held down wages (interestingly, "wages" is a term derived from the same Latin root as the term for "waging war"), just as they have depressed the paychecks of nearly every salaried employee not working on the top floor. The result is that the prosperity generated by the many has been stolen by the few, especially by the million or so richest Americans, and there is a widening chasm between the wealth of these elites and the well-being of the great majority. Writing about the last time such a separation was made in our country, in the 1920s and 1930s, Woody Guthrie penned these lyrics: "As through this world I've traveled/I've seen lots of funny men/Some'll rob you with a six gun/Some with a fountain pen."

Today's fountain-pen bandits do their finagling in the name of "global competitiveness," but they are merely feeding us globaloney when they say that. These privileged and powerful ones have severed themselves from the common good (1) because they can, and (2) because there is a "screw you" imperative built into the corporate system that commands top managers to produce as much money as possible for the big shareholders (including the top managers) as quickly as possible, no matter who or what has to be run over.

While the bandits are loose in the private sector, plundering the middle class by day and stealing back at night to their gated-and-guarded compounds of wealth, those running our public sector have adopted the same pernicious philosophy of "I got mine, you get yours." Health care for all? Hey, Newt and his gang of money-corrupted cutthroats shout, buy yourself a medical savings account. Universal education, the notion that our communities and our nation will be enriched if we are smart enough to provide a system of quality public schools for all? Hey, the ideologues of separation shout, we want to take our tax dollars out of the common pool and fund exclusive academies for our kids, the hell with your ragamuffins. Old-age protection through a shared retirement fund? Hey, shout Wall Street's minions in Congress and the White House, sever the bonds of common responsibility and let everyone bet the farm by putting their social security money in a dazzling array of Wall Street schemes and scams.

The disuniting of our society is the result of a historic failure of our political system. The workaday majority, the folks who believe in America's concept of the common good, have no home in the two major parties, both of which have been molded into branches of one Corporate Party that is pushing hard for disunification. There is no political redress in Washington (or in state capitals) for working people, inner city and rural residents, old folks and children, small farmers, the poor (working and otherwise)—no redress for the middle class and less fortunate.

Bill Clinton himself is a quisling in the class war. Ironically, Clinton's second inauguration was held on the national holiday established to honor Martin Luther King Jr., and of course Clinton invoked the King name to demonstrate his concern for racial tolerance. But Dr. King was not only a martyr for minorities and the racial cause—his agenda was one of economic power for the working class. He was assassinated in Memphis, where he had gone in 1968 not to register African-Americans to vote, but to support striking garbage workers and to organize white and black folks alike to join him in the massive "Poor People's March" he had scheduled for Washington—a trip he never got to take. In the eyes of the country's power brokers, King had moved from being an acceptable missionary of civil rights to being an unacceptable meddler in economic rights. Clinton did not mention this richly heroic side of Dr. King, much less invoke any of the great crusader's populist message. No surprise, given Clinton's essential Republicanism. Indeed, had King been alive on January 20, 1997, he most certainly would have shunned the inauguration of this pusillanimous President, probably instead organizing a march of the "Other Americans" for a show of strength against a "Democratic" administration that is owned and managed by the comfortable and the corporate.

Clinton-the-Democrat campaigned in '92 as the candidate of economic justice, but he has governed as the President of economic privilege. Again in the '96 campaign he sounded populist themes, jumping the Republicans for their assault on Medicare, but now he has met in back rooms with the GOP to plot a bipartisan raid not only on Medicare, but also on Medicaid and social security. Ironically, he agreed to this Republican assault on the elderly and the poor the same week he participated in the dedication of the Franklin and Eleanor Roosevelt memorial in Washington, thus setting a new Olympic record in political hypocrisy.

Having Clinton as a Democratic President is like getting bitten by your pet dog—the bite will heal, but you never feel the same about that dog again.

Have heart, though, for a change is coming. It's coming because it must, just as it has had to come periodically in our history when economic elites have set themselves too far above the rest of us. "The fruits of the toil of millions are boldly stolen to build up colossal fortunes for a few, unprecedented in the history of mankind," proclaimed the preamble to the 1892 platform of the People's Party, rallying the Populist movement to a historic reshaping of America's political landscape. "We have witnessed for more than a quarter of a century the struggles of the two great political parties for power and plunder," the platform raged, "while grievous wrongs have been inflicted upon the suffering people. We charge that the controlling influences dominating both these parties have permitted the existing dreadful conditions to develop without serious effort to prevent or restrain them. Neither do they now promise us any substantial reform. . . . They propose to sacrifice our homes, lives and children on the altar of mammon; to destroy the multitude in order to secure corruption funds from the millionaires."

No need to appoint a drafting committee to put forward the particulars of today's populist manifesto—just fax this one across the country.

A change is coming because there is no party representing the worker majority of our country, creating a political vacuum that a civil society cannot survive. The vacuum exists because today's national Democratic Party is immobilized by its corporate sponsors and its craven leadership, no longer having either the inclination or the political freedom to nod to that majority, much less represent it.

The good news is that a change is coming because it *is* coming, already organizing and gaining strength at the country's grassroots. One vital development is that organized labor has shaken itself awake and is on the move again, with savvy and aggressive new leadership at all levels, adding energy, focus, and resources to the long-term process of building a new political path. There are groups like ACORN, Citizen Action, the PIRGs (Public Interest Research Groups), and Clean Water Action working door-to-door in thousands of neighborhoods all across the country, organizing and training "just folks" to fight back. There are now serious-minded third parties that are beginning to crack the lock that the Democrat and Republican parties have had on the system—a smart and well-organized group called the New Party, for example, already has more dues-paying members than the national Democratic Party; it is running candidates for school boards, city councils, state legislatures, and other local offices in dozens of cities and towns (of 185 campaigns it has run so far, it has won 126); it has teamed with ACORN and others to win "living-wage" campaigns in cities across the country, effectively raising the local minimum wage to as high as $7.50 an hour; and it has had the brains and chutzpah to challenge the election laws that presently rig the political system for the perpetuation of the two major parties. In addition there are groups like the Labor Party and the Alliance for Democracy that are not yet running candidates, but are working throughout the country to inform, educate, and activate citizens to take charge. Also, an essential network of communications already is in place for this burgeoning progressive political movement, including magazines and papers, hundreds of important Internet sites, radio talk shows, newsletters, cable television broadcasts, videos, cartoonists, and musicians.

Most significantly, though, people themselves are ready to fight back and are actively in search of a new politics that enlists them, that stands unequivocally for working families against the bandits, that rejects trickle-down policies in favor of percolate-up economics, and that is centered on that philosophical notion I first learned from my ol' daddy, which bears repeating here: *Everyone does better when everyone does better.*

Daddy also would have agreed with the thought of "fellow

philosopher" Lao-tzu, the sixth-century Chinese founder of Taoism, who observed: "To lead the people, walk behind them." You have to walk pretty briskly, though, for people today are moving way out front of their so-called leadership. Three examples:

1. In the fall of '97, the full line-up of establishmentarians was rolled out to make sure that one of the power tools of corporate globalization was put in the hands of Bill Clinton. Called "Fast Track," this legislation would give the President (acting on behalf of the corporations) the antidemocratic authority to negotiate more NAFTA-like trade deals around the world, then ram them through Congress with no real debate and no chance to amend them. Corporate chieftains, the national media, every Republican congressional leader, all living ex-presidents, the memory of all dead presidents, Henry Kissinger, the entire army of K Street lobbyists, Old Glory, the Pentagon, and six of the Seven Dwarfs were marshaled to say to Congress that "Fast Track" *must, absolutely must, absolutely must with mustard on it* be approved. But it wasn't. To the shock and amazement of the Beltway Crowd, the *people* rebelled. Oh, them. Just average Americans, who were assumed to be disinterested or dumb about trade matters, had already swallowed more globaloney than they could stand, so they raised a righteous ruckus against "Fast Track" that even brain-dead members of Congress could not ignore. I know the depth of feeling that people have about globalization, because for two months leading up to this vote in Washington I talked with them about it every day on my call-in radio program. Working with Citizens Trade Campaign, we targeted two to four undecided lawmakers each broadcast day, gave out the names of the staffers handling the "Fast Track" issue in those offices, offered an 800 number for the Capitol switchboard, and unleashed the people on their Congress-critters. Not for nothing are our listeners known as "Radio-Active"—they melted the phone lines to the offices we targeted. One of our Georgia listeners reported the day before the scheduled vote that he had called "all 535 members of the House and

Senate, plus Al Gore." He added, "Thank God for 800 numbers." The public outcry was so great that Clinton and Gingrich kept delaying the vote for several days while they browbeat, begged, and all but bribed members to support their dog of a bill. But finally they had to surrender, pulling "Fast Track" from the floor to avoid the humiliation of having such a bold NO slapped across the establishment's face. From that day on, the balance of power on globalization issues switched from the elites to the people, and no longer will Congress just rubber-stamp any trade scam industry wants.

2. The absolute corruption of our political system by corporate money is a nonissue, according to Washington insiders, because people don't care, and besides, money is a free speech issue. Free speech? Holy 1776! I don't recall the *Federalist Papers* explaining that the First Amendment's reference to "speech" means that the voice of Amway Corporation, amplified by a cool $1 million contribution to the Republican Party by Amway's president in April 1997, should be so loud that a few months later the Republican Congress would slyly slip Provision C, Section XI into the 1997 Budget Act, producing a $280 million tax windfall requested by Amway lobbyists. That is not free speech, it's bought-and-paid-for speech. *Caution, startling numbers ahead:* Barely 10 million of us 260 million Americans give *any* money to *any* congressional candidate, presidential campaign, or national party in an election year; only one-fourth of 1 percent of us give as much as $200 in an election year; this leaves a teensy-tinsy one tenth of 1 percent of us (235,000 people) who are the $1,000-and-up political donors; since it takes at least $1,000 to get noticed as a political contributor in Washington, this money system of allocating "speech" leaves 99.9 percent of us voiceless. Yet the media pundits and the political operatives think we don't care? They should go to Arizona, Maine, Massachusetts, Michigan, Missouri, North Carolina, Vermont, Wisconsin, or any of two dozen other states where citizens care so much they have either passed or are presently pushing major cam-

paign finance reforms that remove private money from the public business of elections. People know that Washington won't lift this load, so they have to do it for themselves at the state and local levels. In practically every place that has had a vote on reform, folks have not merely cared, they have rallied, organized, and won. In my town of Austin, for example, a reform initiative was put on the ballot through the tenacious grassroots efforts of a coalition with the best political moniker I ever came across: "Austinites for a Little Less Corruption." The entire power structure, from city officials to the media, came down hard on this proposal, trying to submarine it, ridiculing it, lying about it, and generally foaming at the mouth about it for weeks. But on Election Day 1997, a resounding 73 percent of Austinites said Yes! to it.

3. While Washington turns a blind eye to sweatshop products pouring into our country bearing brand-name labels, people like Ed Boyle and Tico Almeida are doing something to stop it. Tico, a student at Duke University, noticed that his school was selling everything from basketball jerseys to electronic games with the Duke "Blue Devils" logo emblazoned on them—$20 million worth of goods a year, produced under contracts the school has with Nike, Russell Athletic wear, and hundreds of other manufacturers. But wait, thought Tico, Nike is a notorious exploiter of teenage Asian girls, effectively holding them as indentured workers in sweatshops. Forming a campus group called Students Against Sweatshops, he began in September 1997 to poke, prod, organize, and protest. Within six months, administration officials had to concede that Duke's credibility as an institution that promotes high moral standards for youth was being tarnished by the pell-mell pursuit of these product sales. "We cannot tolerate having the sweat and tears of abused and exploited workers mixed with the fabric of the products which bear our marks," said the utterly chagrined director of trademark licensing for the school. But Students Against Sweatshops wanted more than a baring of the anguished academic soul, they wanted results. Like a num-

ber of corporations (including even Nike), Duke put forth a "code of conduct," nobly declaring that licensees must refrain from using child labor, must maintain safe workplaces, must pay at least the minimum wage of the countries where they operate, and so forth and so on. But, as workers worldwide know, these codes are worthless if not enforced, and this is where the students achieved a major breakthrough. Duke will now require that its licensees provide the precise location of all factories making products for the university, and that they allow independent monitors (paid for by Duke) to inspect these plants at any time, including unrestricted access to the workers and to human rights groups that are familiar with conditions in the plants. No inspection, no license. Duke is a private institution, but even our public institutions—from state prison systems to the local library—use our tax dollars to buy many goods made by sweatshop labor. If Ed Boyle has his way, this will stop. As mayor of North Olmsted, Ohio, a blue-collar suburb of Cleveland, he came up with a way to ban city purchases of sweatshop products—a 1997 North Olmsted ordinance requires vendors to sign a legal document swearing that each and every product sold to the city is free of child or forced labor, is produced by employees who are free to form unions of their choice, is made in a safe and healthy workplace, and so on. If a vendor won't comply, or lies—no sale. "I had my worries at first," the mayor told writer John Nichols. "We're a small city and I didn't know if we were the ones to take the lead on this. But it's worked. We're reaching Americans who think like I do—that this country is supposed to be against exploitation. If there's one thing that the government of this country, from Washington right down to North Olmsted, ought to be against, it's sweatshops."

There is no need to "create" a progressive movement, because it already exists in the hearts and minds of America's ordinary folks. Instead the chore of progressive strategists and organizers is to connect these folks nationally and help them build an independent political mechanism that frees them from reliance on either branch of today's corporate, one-party structure.

There was a moving company here in Austin that had an advertising slogan I liked: "If we can get it loose, we can move it." That's the spirit! If we give people a means to get loose from the old politics, they'll get moving—and they'll take America back.

INDEX

Perennial

Books by Jim Hightower:

THERE'S NOTHING IN THE MIDDLE OF THE ROAD
BUT YELLOW STRIPES AND DEAD ARMADILLOS
A Work of Political Subversion
ISBN 0-06-092949-9

Hightower takes aim at those bedrock institutions that drive the economic and cultural life of the country. In this lucid, viciously funny, downright refreshing book, Hightower argues that government, the media and corporate conglomerates have put us smack-dab in the middle of the mess we're in today. Revised with a new introduction.

"If you don't read another book about what's wrong with this country for the rest of your life, read this one. I think it's the best and most important book about our public life I've read in years."—Molly Ivins

IF THE GODS HAD MEANT US TO VOTE,
THEY'D HAVE GIVEN US CANDIDATES
More Political Subversion from Jim Hightower (Revised Edition)
ISBN 0-06-093209-0

A piece of Hightower's mind on "Candidates 2000" and the whole stinking political mess. Fully revised and updated with election results and new material, this new book offers a plain-talking, name-naming, podium-pounding, point-them-in-the-right-direction, populist prescription for 2000.

"Forget all the political insider books about America's elections, here's the outsider book that tells you what's really going on in the 2000 election, why, and what we have to do to make politics work for us again."

—William Greider